ACS SYMPOSIUM SERIES **670**

Supercritical Fluids

Extraction and Pollution Prevention

Martin A. Abraham, EDITOR

University of Toledo

Aydin K. Sunol, EDITOR

University of South Florida

Developed from a symposium sponsored by the
Division of Industrial and Engineering Chemistry, Inc.,
at the 211th National Meeting of the American Chemical Society,
New Orleans, Louisiana,
March 24–28, 1996

American Chemical Society, Washington, DC

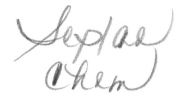

Library of Congress Cataloging-in-Publication Data

Supercritical fluids: extraction and pollution prevention / Martin A. Abraham, editor, Aydin K. Sunol, editor.

p.   cm.—(ACS symposium series, ISSN 0097–6156; 670)

"Developed from a symposium sponsored by the Division of Industrial and Engineering Chemistry, Inc., at the 211th National Meeting of the American Chemical Society,New Orleans, Louisiana, March 24–28, 1996."

Includes bibliographical references and indexes.

ISBN 0–8412–3517–1

1. Supercritical fluids—Congresses.

I.Abraham, Martin A., 1961– . II. Sunol, Aydin Kemal, 1953– . III. American Chemical Society. Division of Industrial and Engineering Chemistry. IV. American Chemical Society. Meeting (211th: 1996: New Orleans, La.) V. Series.

TP156.E8S847   1997
660′.284248—dc21   97–29085
                                                          CIP

This book is printed on acid-free, recycled paper.

PRINTED IN THE UNITED STATES OF AMERICA

Foreword

THE ACS SYMPOSIUM SERIES was first published in 1974 to provide a mechanism for publishing symposia quickly in book form. The purpose of the series is to publish timely, comprehensive books developed from ACS sponsored symposia based on current scientific research. Occasionally, books are developed from symposia sponsored by other organizations when the topic is of keen interest to the chemistry audience.

Before agreeing to publish a book, the proposed table of contents is reviewed for appropriate and comprehensive coverage and for interest to the audience. Some papers may be excluded in order to better focus the book; others may be added to provide comprehensiveness. When appropriate, overview or introductory chapters are added. Drafts of chapters are peer-reviewed prior to final acceptance or rejection, and manuscripts are prepared in camera-ready format.

As a rule, only original research papers and original review papers are included in the volumes. Verbatim reproductions of previously published papers are not accepted.

ACS BOOKS DEPARTMENT

Contents

EXTRACTION AND CHROMATOGRAPHY

ENVIRONMENTAL APPLICATIONS

Preface

THE USE OF SUPERCRITICAL FLUIDS FOR EXTRACTION is growing rapidly, with new applications being developed almost daily. Enhancements in techniques, analytical capabilities, and equipment are leading to greater use of supercritical fluids in areas such as polymers/materials processing, thermodynamics, recovery of natural materials, chromatography, and environmental applications. This book was developed from a symposium consisting of a series of five sessions titled "Supercritical Extraction", presented at the 211th National Meeting of the American Chemical Society sponsored by the ACS Division of Industrial and Engineering Chemistry Inc., New Orleans, Louisiana, March 24–28, 1996. The symposium brought together researchers from around the world to present an overview of state-of-the-art applications of supercritical fluid extraction.

The symposium was divided into sessions according to the applications area. To provide a solid introduction into each of these areas, each session consisted of one invited paper and then six contributed papers. The same structure has been maintained within this book, although the materials processing and thermodynamics sections have been combined. Each section is introduced by a lead paper, authored by the leading experts in specific areas. These papers describe both the new research being conducted by the authors, as well as a review of the current state of science within the particular application area.

Contributions from many of the top research laboratories throughout the world are included in this book which represents one of the first international compilations on supercritical fluid extraction. This book should be of interest to both academic and industrial researchers. Many of the papers describe fundamental developments, and yet remain practical to the engineer or scientist needing to apply supercritical fluid extraction to his/her job. It is carefully placed on the precipice between fundamental and applied research.

Acknowledgments

The editors thank all those who have contributed to bringing about this book including the authors, who have contributed their papers to this volume, and the reviewers, who worked hard to ensure the quality of the contributions. We also thank the ACS Division of Industrial and Engineering Chemistry, Inc. that gave us the opportunity to organize these sessions; Anne Wilson and Cheryl Shanks, the editors at ACS Books, who have patiently resigned themselves to our delays and missed deadlines; and our secretarial staff and colleagues at The University of Tulsa, the University of Toledo, and the University of South Florida. Finally, many thanks go to our families, who have sacrificed so that we could put the time and effort into making this book possible.

MARTIN A. ABRAHAM
Department of Chemical Engineering
University of Toledo
Toledo, OH 43606–3390

AYDIN K. SUNOL
Department of Chemical Engineering
University of South Florida
Tampa, FL 33620

July 9, 1997

Macromolecules and Thermodynamics

Chapter 1

Miscibility and Phase Separation of Polymers in Near- and Supercritical Fluids

Erdogan Kiran and Wenhao Zhuang

Department of Chemical Engineering, Jenness Hall, University of Maine, Orono, ME 04469-5737

The factors that influence high-pressure miscibility and phase separation of polymers in near- and supercritical fluids are reviewed. These are discussed with an emphasis on their significance in a number of applications including but not limited to polymer formation, modifications, processing, and recycling. Methodologies associated with kinetics of phase separation are discussed. Pressure-induced phase separation (PIPS), its dependence on quench depth, and the consequences of penetration into the metastable (nucleation and growth) or unstable (spinodal decomposition) regions are presented. The examples are drawn primarily from research conducted at the University of Maine.

In the *supercritical fluid state,* the distinction between a liquid and a gas disappears and many properties such as density, viscosity, and diffusivity become adjustable. A wide range of properties from *gas-like* to *liquid-like* become accessible by simple manipulations of pressure and temperature without entering the two-phase region of the fluid. Depending upon the pressure or density, different solvent characteristics become accessible by the same fluid. As a result, the same fluid can be fine-tuned to behave , for example as a specific solvent or non-solvent for a particular substance. Such adjustable properties make supercritical fluids very attractive as process fluids or reaction media for many industrial applications [1-3]. A particularly important area is the polymer industry.

More than 30 million tons of various polymers are produced yearly in the United States which find use in diverse applications ranging from common household appliances and toys, to highly specialized materials used in electronics or those used for medical applications [4]. Among the various commodity polymers, polyethylene, polypropylene, poly(vinyl chloride), polyesters, and phenolics are produced in the greatest amounts. Indeed, different forms of polyethylene (low and high density PE) alone amount to more than one third of the yearly polymer production. Many of the specialized application areas require the use of copolymers or special blends to achieve the desired material properties.

At the University of Maine, we maintain a broad perspective for high-pressure processing and utilization of supercritical fluids for polymer applications [5]. We have been evaluating these fluids for various applications in the formation, modification, processing, recovery and recycling of polymers [5-16]. Advances in these areas greatly depend on our understanding of miscibility and phase separation of polymers at high pressures in near and supercritical fluids, which constitutes the subject matter of this chapter.

This chapter is structured in four parts. Part I provides a brief account of the relative nature of the notion of high-pressure, and highlights the key features of supercritical fluids with an emphasis on binary fluid mixtures. Part II summarizes some basic concepts related to polymers, their formation, properties and processing, and in so doing attempts to provide a broad-base rationale for the importance of supercritical fluids and their utilization. Parts III and IV are devoted to topics related to high-pressure miscibility and phase separation, respectively. A special emphasis is placed on pressure-induced phase separation (PIPS).

Part I. High-Pressure Scale and Supercritical Fluids

Pressure has been recognized as an important parameter in polymer industry for many years. In fact, the free-radical polymerization of ethylene to produce low density polyethylene is conducted at pressures near 2000 bar [17]. Here ethylene functions both as a reactant and a solvent, and the conditions are supercritical for ethylene. However, over the past decade, the scope of polymer applications where supercritical fluids can be used has expanded greatly and is no longer limited to polymerizations. Even for polymerization, the current drive is more on exploring supercritical fluids for their use as general solvent media (i.e., the fluid being different than the monomer) which may be environmentally more acceptable [5, 16]. In this respect, a particularly important development in recent years has been the methodologies that are developed to permit the use of carbon dioxide as polymerization medium [18-20].

"High-pressure" is a relative term, and it is important to appreciate the pressure levels in terms of some easy-to-associate reference points. Table I provides such a pressure scale. For a surface scientist, even 1 bar would be considered high pressure. From the perspective of a diver, 60 bar is rather crucial since at higher pressures human central nervous system is known to break down [21]. This is in contrast to fish species which survive in deep oceans at pressures near 1,000 bar. For example, sharkes function from sea level to extreme depths [22]. An interesting factor here is the effect of pressure on protein (a biopolymer) and its chain-folding characteristics in humans versus other animals. The extreme pressures and also high temperatures that prevail inside earth, where water and carbon dioxide are both supercritical, highlights the importance of hydrothermal fluids in geological processes, particularly in the formation of minerals, gems and their tansformations

[23, 24]. Carbon dioxide and water are two fluids which are environmentally acceptable, and become supercritical at reasonable pressures.

Table I. Pressure Scale

Pressure (bar)	Characteristic system
10^{-12}	pressure in a high-vacuum chamber - surface science studies
1	50-65 km of air column acting on a person at a beach
1.5	pressure cooker in the kitchen
2.0	air pressure in a car tire
50	pressure exerted by a lady in stiletto heels
60	break down of central nervous system for humans
73.8	critical pressure of carbon dioxide
221.2	critical pressure of water
1,000	pressure at the bottom of the ocean (at 10,000 m)
2,000	low-density polyethylene production
10,000	pressure below the crust of earth at 30 km
100,000	pressure synthetic diamond are produced
1,000,000	pressure at the center of the earth at 3000 km

It is important to note that supercritical fluid processing does not necessarily mean extreme pressures. In fact, many fluids become supercritical at relatively low pressures, typically below 100 bar [25]. Majority of organic solvents become supercritical at pressures below 50 bar. Water is one of the few substances with a critical pressure greater than 100 bar . On the other hand, for critical temperatures, there is a wider range of values displayed depending upon the chemical structure of the fluid and its boiling point. For example, the critical temperatures for carbon dioxide, pentane, and toluene are 31, 196.7 and 318.6 °C, respectively. For water critical temperature (374.1 °C) is even higher.

At the vapor-liquid critical point, the distinction between a gas and a liquid disappears. At supercritical conditions, the substance is a one-phase homogenous fluid. Its properties can be continuously changed between those of a gas and a liquid by changing either pressure or temperature without entering the two-phase regions. Clearly, by changing the temperature and the pressure, the properties can be changed in the liquid or gaseous state also, but such changes are much smaller than changes that are achieved in the supercritical state. The properties of supercritical fluids are fine-tunable by simple manipulations of the pressure or the fluid density to assume an extremely wide-range of values and therefore, are very useful as process or processing fluids.

Carbon dioxide with its low critical temperature and environmentally acceptable nature is highly desirable as a supercritical processing fluid. However, for many polymer applications where complete miscibility may be desirable, carbon dioxide is of limited utility and therefore mixtures are often considered. Mixtures of carbon dioxide with many organic solvents such as butane, pentane, toluene, cyclohexane, methanol, ethanol display continuous critical lines. These mixtures, depending upon the composition of the mixture display critical temperatures that are in between the critical temperatures of the two pure fluids, while their critical pressures pass through a maximum. This is demonstrated in Figure 1a for mixtures of carbon dioxide + pentane. Figure 1b shows the variation of the density for these mixtures with pressure for selected compositions covering sub-critical and supercritical temperatures at pressures above the critical pressure for all mixtures. This figure highlights another interesting feature of carbon dioxide which, even though a gas under normal conditions, when compressed spans much higher densities than a fluid like pentane which is a liquid under normal conditions. Density which is a function of temperature and pressure is an important parameter which determines the properties of the fluid and, as shown in this figure, can be conveniently adjusted (for a specific end-use) also by employing binary fluid mixtures. Density tuning by using fluid mixtures is key also to a new concept on density-modulated levitation processing [16].

At the University of Maine, we work extensively with binary fluid mixtures of carbon dioxide with organic solvents to (a) achieve lower critical temperatures (and thus lower processing temperatures) than that for the organic solvent, (b) to introduce selectivity and/or reactivity that carbon dioxide alone cannot provide, (c) to modulate the fluid density and its properties in some unique ways, and (d) to identify alternative processing fluids with reduced amount of undesirable solvents. We investigate pressures up to 1000 bar (representative of the bottom of the ocean).

Part II. Basic Concepts Related to Polymers and Significance of Supercritical Fluids Processing

In order to demonstrate the significance of supercritical fluids and their adjustable properties in diverse polymer applications, it is important to review some of the characteristics features of polymeric systems that pertain to polymer formation, properties and processing.

Polymer Formation. Polymers are long-chain molecules that are produced by either chain-addition or step-growth mechanisms. These polymerization processes, whatever the mechanism, are random processes. As a result, polymer chains growth to different extend, leading to a distribution of chain lengths or molecular weights. In co-polymerizations involving two or more monomers, not only chains of different length but chains with different monomer sequence distributions are formed. Typically, the end product from bulk polymerization reactions, in addition to the

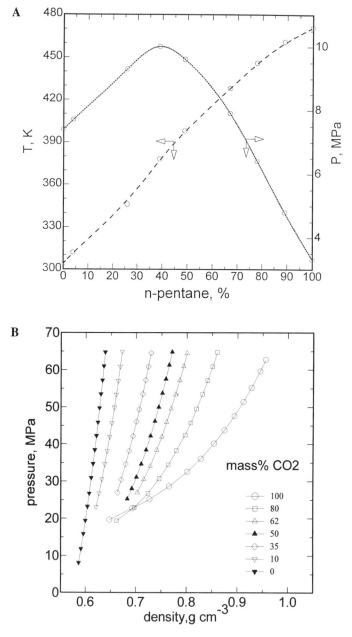

Figure 1. A (Top). Critical temperature and pressure of binary mixtures of carbon dioxide and pentane. [Data from ref. 41]. **B (Bottom).** Variation of the density with pressure for mixtures of carbon dioxide and pentane. T = 348 K. Compositions are on mass basis. [Data from ref. 42].

polymer chains of different lengths or chain composition, contain unreacted residual monomer(s), residual initiator and catalyst molecules, and low molecular weight oligomers. If the polymerization is conducted by solution/precipitation, dispersion or emulsion polymerization procedures, the end product may also contain solvent residue, and additives used as emulsifiers. Extraction of residual monomer, solvent and other non-polymeric constituents and impurities, and the fractionation of polymers with respect to molecular weight or chain composition are among the most direct application areas for supercritical fluids. Sequential or selective separations are achieved by either the pressure, the temperature, or the solvent tuning of the extraction fluid, or by a combination of these parameters. A major activity area is the use of supercritical fluids as polymerization media (a) to replace conventional organic solvents, (b) to control the molecular weight and molecular weight distribution, and (c) to control the microstructure and morphology of the polymer formed. Again the pressure, fluid composition and density are used as the tuning parameters. Recently, density tuning has been used in a unique way for levitation of polymers formed in supercritical fluid media [16]. A special case of polymerization is the curing process in thermosetting polymer systems in which use of supercritical fluids may lower the viscosity, and the glass transition temperature, thereby delaying vitrification and enhancing conversion or degree of cure [26]. The fluid can then be used to remove (extract) unreacted monomer or by products.

     Property Modification. Depending upon the chemical structure, chain configurations, and the molecular weight, polymers display different properties. In the solid state, polymers may be amorphous, crystalline or semi-crystalline. These different morphological forms are characterized by the glass transition, melting transition, or both the glass and the melting transition temperatures, respectively. Below the glass transition temperature a polymer is glassy and rigid with a high modulus. The glass transition is intimately related to chain mobility and free-volume, and as such, in the presence of small molecules or diluents, the glass transition temperature can be significantly lowered. A major application area for supercritical fluids involves lowering the glass transition temperature by dissolution of supercritical fluids in the polymer to bring about softening to facilitate impregnation with additives (such as drugs for controlled-release formulations), dyes (as in textile dying), or with other monomers (that are then polymerized in-situ) or polymers (for producing micro-blends). The solubility and the amount of fluid in the polymer and the extend of impregnation can be easily altered by changing the tuning parameters such as pressure. Clearly, the extend of modifications can be restricted to the surface layers, or may be permitted to be more extensive. Lowering of glass transition is also important in extraction of constituents from a given polymer material which are important for purification or recycling applications. Lowering the glass transition temperature introduces also new possibilities for crystallization of semicrystalline polymers. For polymers which can crystallize, crystalline morphology depends whether the crystals are formed from solution or melt, and normally also on the crystallization temperature, i.e. on the degree of under-cooling (with respect to the melting temperature) that is imposed. For semi-crystalline polymers, since rate of crystallization decreases drastically as the glass transition temperature is approached, lowering the glass transition temperature in the presence of supercritical fluids, also moves the practical range of crystallization temperatures to lower temperatures. More generally, the use of supercritical fluids introduces pressure as an important parameter to induce crystallization and influence final morphology.

Processing. Final processing of polymers to produce end-use products often involve processing in the molten state (as in the case of extrusion or injection molding operations), or from solutions (as in the case of spinning for fiber formation, or atomization for particle formation, or spraying for coatings). In these operations, viscosity of the melt or the solution, and the rate of removal of the solvent from the end products are important parameters. For these applications, supercritical fluids can be used to reduce the viscosity of the molten polymers, or form polymer solutions with lower viscosities than in ordinary liquids. Expansion from supercritical solutions with pressure-induced phase separation provides new methodologies for producing particles, coatings, fibers, foams, membranes and other porous structures. Impregnation of microporous substrates with polymer solutions in a supercritical fluid followed by pressure-induced phase separation offers alternative approaches to produce novel composites materials. A special application area would be improved penetration of pre-polymers and/or monomers into swollen matrices, or fibrous networks for the manufacture of high-tech composites.

Recycling. Because of environmental concerns, there is much interest in recycling of polymeric materials. Plastics represent about 8 % by mass (18 percent by volume) of the more than 200 million tons of municipal solid waste generated each year in the United States [15, 27]. Currently, only a small fraction (3.5 %) of the plastic waste is being recycled. Majority of the recycling effort with polymers is in the recovery of polymeric materials from containers and packaging, the largest item being PET from the soda bottles. Beyond beverage bottle cycling, progress has been limited due to a number of difficulties. The materials recovered cannot automatically go back to the same end-uses. Before any reprocessing, they must be separated into relatively pure forms. For example, it takes only about 10 ppm of PVC containers to contaminate a load of PET [28]. Solid contaminants, such as pigments and metals cause problems with color which limit the utilization of recycled materials. Simple separation techniques such as floatation, or sorting fails when multi-layered co-extruded polymers, or blends are involved. Difficult-to-separate mixed plastics are often processed into low-grade and highly contaminated products such as "plastic lumber" used in such applications as park benches and boat rocks. The general practice to eliminate contamination is a drastic one which involves depolymerization of the plastic waste into monomers which are then purified and repolymerized (i.e., tertiary recycling). A dissolution process involving multiplicity of solvents such as xylene and THF has been recently proposed requiring however rigorous solvent recovery stage and de-volatilization of the polymer since most of the solvents used are high-boiling liquids [29]. It is in this area that supercritical fluids with their adjustable penetration and dissolving power offer convenient alternatives for solvent-based extraction and recovery of constituents from co-mingled plastics [5, 15]. Wide range of selectivities can be attained with pressure, or temperature tuning using a single fluid system. Extracted polymers can be recovered by relatively simple pressure reduction steps. There is the potential that in a single process, a given mixed waste, without strict pre-sorting, can be fractionated or purified into usable forms.

The forgoing discussion, by no means comprehensive in coverage, was meant to emphasize the great potential and the suitability of supercritical fluids for polymer applications. To recapitulate, the primary application areas for supercritical fluids in the polymer industry encompass an extremely large range including but not limited to (a) polymer formation, (b) purification, (c) fractionation, (d) property modification and processing, and (e) recycling and recovery. For all these applications, the fundamental information that is needed is clearly centered around those factors that influence miscibility and phase separation of polymers in supercritical fluids at high pressures.

Part III. Miscibility of Polymers in Near- and Supercritical Fluids

The following are the key factors that determine the miscibility of polymers in a given fluid system:

polymer type
molecular weight and molecular weight distribution
nature of the solvent
polymer concentration
temperature
pressure.

From a thermodynamic perspective, for complete miscibility, the free energy change upon mixing must be negative ($\Delta G_m < 0$) and for binary mixtures, its second derivative with respect to composition must remain positive ($[\partial^2 \Delta G_m / \partial \phi^2]_{T, P} > 0$). In partially miscible systems, even though ΔG_m is negative for all compositions, it shows an upward bend in a specific range of compositions identified by say ϕ_{Ib} to ϕ_{IIb} [Figure 2a] where the first derivatives (i.e., the chemical potentials) become identical (i.e., a common tangent can be drawn to the Gibbs function as shown in the figure), and the system phase separates into two phases (a polymer lean phase and a polymer rich phase). These compositions define the equilibrium compositions or the binodal points. In the composition range inside the binodal, two other compositions ϕ_{Is} and ϕ_{IIs} known as the spinodal points can be identified where the second derivatives become zero. These correspond to the inflection points in the ΔG_m function. Within the spinodal range, the second derivative becomes negative (Figure 2a, bottom curve) and the system becomes thermodynamically unstable leading to spontaneous phase separation known as the spinodal decomposition. The composition range between the binodal and the spinodal is a metastable region for the system. With changes in the temperature or the pressure, the shape of the Gibbs function, and consequently the compositions corresponding to binodals and the spinodals can be changed, and the regions of miscibility and immiscibility can be altered. This is demonstrated in Figures 2b and 2c. These figures are applicable for system in which either the temperature is

A

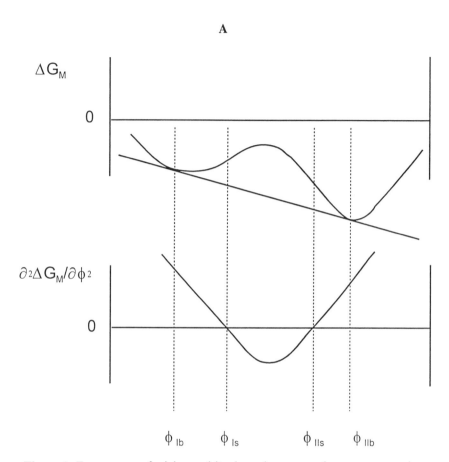

Figure 2. Free energy of mixing and its dependence on polymer concentration and external parameters. **A (Left).** Free energy of mixing for a partially miscible system and the common tangent identifying composition of equal chemical potential ϕ_{Ib} and ϕ_{IIb} (i.e., binodals) (top), and the second derivative of the free energy of mixing identifying the compositions for the spinodal conditions, ϕ_{Is} and ϕ_{IIs}. **B (Center).** Change in free energy of mixing with external parameter, i.e., temperature at constant pressure, identifying the loci of binodal and spinodals and the upper critical solution conditions. Curves d, c, b, a correspond to increasing value of the parameter (i.e., temperature) that is changed. In Figure **C (Right).** curves d, c, b, a represent decreasing values of the parameter changed and demonstrates a system displaying lower critical solution conditions, i.e., a system becoming miscible by lowering for example the temperature.

B

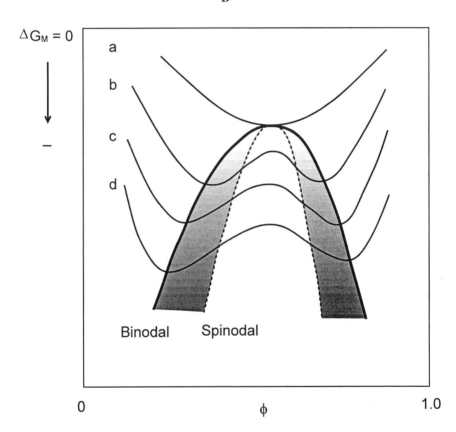

Figure 2. *Continued*

continued on next page

C

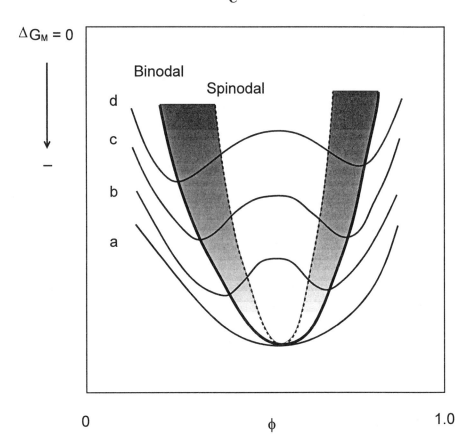

Figure 2. *Continued*

changed (while holding the pressure constant) or the pressure is changed (while holding the temperature constant). The loci of the binodal and the spinodal points define the boundaries for the stable, metastable and the unstable regions of the system. These curves merge, i.e., the binodal and spinodal points become identical, at a specific temperature (holding pressure constant), or at a specific pressure (holding temperature constant) defining a critical solution temperature or pressure and a corresponding critical polymer concentration for the system. The change in the shape of the Gibbs function with the change in temperature (or pressure) and the eventual disappearance of the metastable and unstable regions for either the increasing (curves c, b, a in Figure 2b) or the decreasing (curves c, b, a in Figure 2c) values of the external parameter (i.e., temperature) is demonstrated in Figures 2b and 2c. Beyond the upper or lower critical solution conditions, the homogeneous state is the global minimum of the free energy, and there are no longer local minima leading to metastable or unstable regions.

For polymer systems, it is well known that, at a given pressure, complete miscibility in conventional solvents can take place either upon an increase or a decrease in temperature. Those that become miscible with increasing temperature are characterized by an upper critical solution temperature (UCST), and those becoming miscible with decreasing temperature are characterized by a lower critical solution temperature (LCST). As will be demonstrated in the following sections, polymer solutions in near and supercritical fluids at high pressures, depending upon the system, may display UCST or LCST type behavior. However, holding temperature constant, complete miscibility is almost invariably achieved by increasing pressure, identifying an upper critical solution pressure (UCSP).

A basic task in polymer solution studies is the identification of the characteristic phase boundaries. Experimental approach often involves determination of the cloud points (demixing points) which for truly monodisperse systems would be identical to the binodal boundary. Experimental determination of the spinodal boundary uses extrapolative procedures based on light-scattering measurements.

Miscibility and Demixing Pressures

Demixing pressures are determined using high-pressure variable-volume view cells. In these experiments, the cell is first loaded with the polymer and the solvent(s) to achieve a target concentration, and then the temperature and / or the pressure are changed to achieve complete miscibility. At each temperature, the pressure is then reduced to induce phase separation, and the demixing pressures (cloud points) are noted either visually or by optical measurements. Through such experiments, a series of pressure-temperature curves are generated which define the homogeneous one-phase and the phase-separated regions. Figure 3 is such a plot for solutions of polyethylene ($Mw=16,400$, $Mw/Mn = 1.16$) in n-pentane for different concentrations. For each concentration, the region above each curve represents the homogeneous regions which are entered either by an increase in pressure at a given temperature, or by a decrease in temperature at a given pressure. This system displays LCST

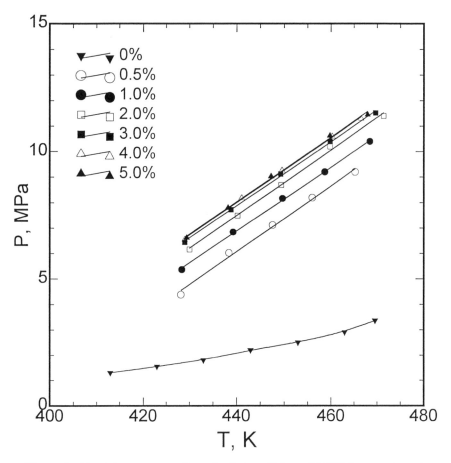

Figure 3. Demixing pressures for polyethylene (Mw = 16,400) in n-pentane at
different polymer concentrations (mass %). The region above each curve is the
completely miscible, homogeneous one-phase region. The curve corresponding
to pure pentane (0 % polymer) is the vapor pressure curve for pentane ending at
its critical point.

behavior. From such plots, by taking constant temperature cuts, pressure-composition (P-x), or by taking constant pressure cuts, temperature-composition (T-x) diagrams can be generated. These are demonstrated in Figures 4a and 4b. In figure 4a homogeneous regions are above each curve (miscibility takes place upon increase in pressure), while in Figure 4b the homogeneous regions are below each curve (miscibility takes place upon decreasing the temperature). Figure 5 shows the P-x diagram for different molecular weight polyethylene samples.

These figures demonstrate features that are common to many polymer solutions. To achieve complete miscibility, higher pressures are needed for higher molecular weight samples. For a given molecular weight sample, demixing pressures increase rapidly at low concentrations, but often becomes less sensitive to concentration over a fairly wide range, and then decreases as the polymer concentration is further increased. Whether the systems display LCST or UCST behavior however strongly depends on the polymer-solvent system. For example, even though solutions of polyethylene in pentane show LCST, in binary mixtures of pentane + carbon dioxide the system behavior changes from that of LCST to UCST with increasing carbon dioxide content of the solvent. This is demonstrated by the change in slope of the demixing pressures as shown in Figures 6a and 6b for the PE samples with molecular weights 16,400 and 108,000 at 5 mass % polymer concentration. Figure 6 demonstrates another important observation, that is carbon dioxide is in general not a good solvent for polymers, and with the addition of carbon dioxide to the system, much higher pressures are needed to achieve complete miscibility. For the polymer with molecular weight 16,400, pressure increases from about 10 MPa in pure pentane to about 70 MPa when the carbon dioxide content of the solvent is inceased to 50 % by mass. At the same concentration, for a polyethylene sample of molecular weight 2150, the demixing pressures increase to 70 MPa at a carbon dioxide content (in the solvent mixture) of 70 % by mass [30], while for a sample with molecular weight of 108,000, as shown in Figure 6b, 70 MPa is reached at 40 % by mass carbon dioxide level. The behavior of 108,000 sample is shown in a computer simulated ternary diagram in Figure 7 at 400 K for two different pressures, 25 and 150 MPa [8]. At 25 MPa, the system can tolerate only up to about 25 % carbon dioxide if complete miscibility must be maintained. The one-phase region expands with increased pressure, or higher carbon dioxide contents are tolerable at higher pressures. As shown in the figure, pressures must be increased to 150 MPa and higher, if the mixture were to contain 50 % carbon dioxide and complete miscibility were to be maintained. It is easy to deduce that to dissolve high molecular weight PE in pure carbon dioxide would require extreme pressures. In fact, there are only a limited number of polymers that can be dissolved to any extent in carbon dioxide. Polydimethylsiloxane is the best known example. In the temperature range 340 to 460 K, molecular weights up to 370 K have been recently shown to be completely miscible in carbon dioxide at pressures below 50 MPa [6]. At lower temperatures however, the demixing pressures show a steep increase. Low molecular weight paraffin wax can be dissolved in carbon dioxide at pressures above 60 MPa. [31].

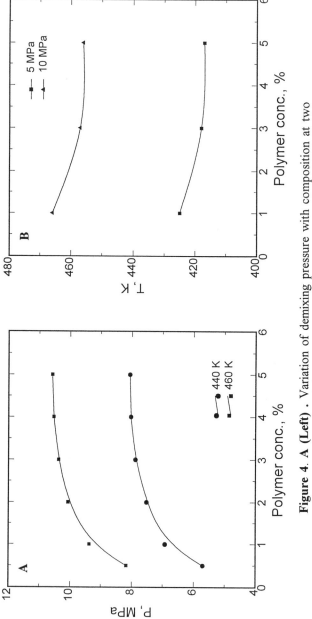

Figure 4. A (Left) . Variation of demixing pressure with composition at two temperatures generated from Figure 3 for the system PE (M = 16,400) + pentane. The region above each curve is the homogeneous region. **B (Right)**. Variation of the demixing temperature with composition at two different pressures generated from Figure 3 for the same system. Here, it is the region below each curve that is the homogeneous region. System undergoes phase separation with an increase in temperature.

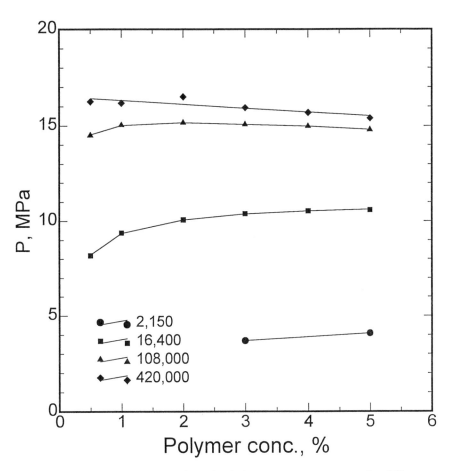

Figure 5. Demixing pressures for polyethylene + pentane system for different molecular weight polymer samples. Demixing pressures increase with molecular weight, and higher pressures are needed to bring about complete miscibility.

Even though in the presence of carbon dioxide the demixing pressures are increased, such binary mixtures offer interesting advantages. Because of different sensitivity is displayed depending upon the molecular weight, or the type of the polymer, carbon dioxide content of the mixture becomes a tuning parameter which can be important in fractionation of a given polymer, or in separation of mixed polymer systems, or can be simply used as a non-solvent to bring about phase separation in the system. A second important consideration is that, as pointed out earlier, the solvent mixture may now be in the supercritical state at the operational temperature and pressures in contrast to the pure organic solvent, which may permit recovery and the recycle of the solvent mixture without vapor-liquid phase separation. A third consideration is the reduction of the amount of organic solvents for environmentally conscience processing options. For these reasons, we have been generating extensive data for ternary systems "polymer + solvent + carbon dioxide" [8, 11, 15, 43, 44]. As a specific example, Figure 8 shows the miscibility boundaries for a number of polymers at a 3 mass % concentration in carbon dioxide (30 %) + pentane (70%) mixtures. A great degree of variation is demonstrated in demixing pressures which can be used for example in selective extractions for recycling of mixed waste [15].

Part IV. Pressure -Induced Phase Separation (PIPS)

Phase separation in polymer solutions can be induced by decreasing the solvent quality through changes in

temperature,
pressure, or
solvent composition.

Hydrodynamics of a system (applied shear) and interfacial tension are also factors that may influence miscibility or phase separation since shear may promote burst of droplets, and coagulation is driven by interfacial tension.

Even though thermally-induced or solvent-induced phase separation are the common methods, pressure-induced phase separation is more important for systems involving near- and supercritical fluids. This is because, whatever the application, pressure reduction is an integral step in any supercritical fluid processing to recover the end-product of interest or the solvent. Equally important is the fact that pressure can also be used to bring about rapid phase separation and lock a polymer-rich phase in a non-equilibrium state to achieve different morphological characteristic or micro heterogenities.

Phase separation can be discussed in terms of phenomena that occur in the early and later stages of the process. In the early stages, concentration fluctuations that occur in the homogeneous one-phase solution lead to the formation of new phases. The later stages involve the coalescence of these macroscopically distinguishable domains. In the early stages of the formation of the new phase, two

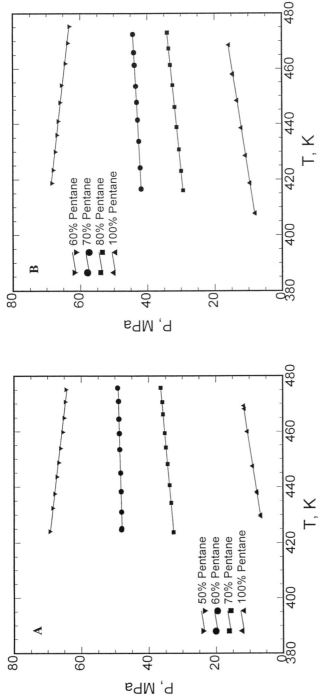

Figure 6. A (Left) Demixing pressures for "polyethylene (M = 16,400) + pentane +carbon dioxide" and **B (Right)** for "polyethylene (M = 108,000) + pentane +carbon dioxide" system as function of pentane content of the solvent mixture. Demixing pressures increase with increasing amount of carbon dioxide and the system behavior shifts from one displaying lower solution temperatures to one displaying upper solution temperatures.

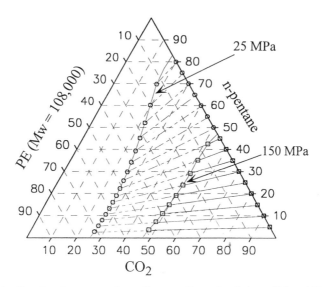

Figure 7. Simulated ternary phase diagram for polyethylene (M = 108,000) + pentane + carbon dioxide system at 400 K. Phase boundaries are demonstrated for 25 and 150 MPa, the regions to the left of each curve being the homogeneous regions.

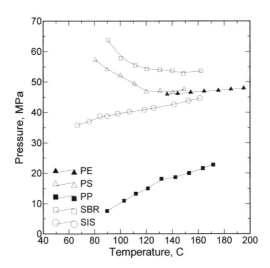

Figure 8. Demixing pressures for "polymer + pentane + carbon dioxide" systems at 3 mass % polymer concentration. Solvent is 30 % by mass carbon dioxide. PE = polyethylene (M_w = 121,000); PP = isotactic polypropylene (M_w = 260,000); PS = polystyrene (M_w = 8,000); SBR = styrene-butadiene block copolymer (M_w = 140, 000; 30 % styrene); SIS = styrene-isoprene block copolymer (M_w = 150,000; 14 % styrene).

different mechanisms are distinguishable. These are known as *Spinodal Decomposition* and *Nucleation and Growth*. Spinodal decomposition occurs when the system is brought into an unstable region. In this region, any concentration fluctuation (infinitesimal or not) leads to a reduction in free energy of the system. Phase separation by nucleation and growth mechanism occurs when the system is brought into a metastable region. Here for infinitesimal concentration fluctuations, the system is stable but for large concentration fluctuations it becomes unstable.

The essential features of pressure-induced phase separation can be described in terms of Figure 9a which is a pressure-composition phase diagram showing the binodal and the spinodal boundaries, and the stable, metastable, and unstable regions. This is an idealized diagram, in that for real solutions, because polymers display polydispersity, the binodal and spinodal curves do not necessarily merge at the maximum of the binodal, but rather shifted to higher concentrations. The region between the binodal and the spinodal are metastable. Solutions in these regions are stable to small fluctuations in compositions, but for large fluctuations undergo demixing. Inside the spinodal envelope, all fluctuations result in a decrease of free energy, and as a result, the solutions are unstable and demixing is spontaneous.

Three paths AB, A'B', and A''B'' represent pressure-induced phase separation in solutions at three different overall compositions upon reducing the pressure from an initial pressure of P_i where the solutions are homogeneous, to a final pressure P_f inside the phase boundary. For the path AB, representing phase separation in the relatively dilute polymer solution, the system enters the metastable region where the new phase formation and growth proceed by nucleation and growth mechanism. Here the polymer-rich phase nucleates and the solvent-rich phase is the continues phase. When equilibrium is reached, the polymer rich-phase has a high polymer concentration of $\phi_{II\ b}$, while the polymer-lean phase has a low polymer concentration $\phi_{I\ b}$. For path A''B'' representing phase separation in a concentrated polymer solution, the system again enters the metastable region, and phase separation proceeds by nucleation and growth mechanism. In contrast to path AB, here the polymer- lean phase nucleates in a polymer-rich phase which is continuous. However, when equilibrium is reached, the final compositions of the two phases would still be the same. Path A'B' corresponds to a solution at its critical concentration, and here upon pressure reduction, the system enters the spinodal region where phase separation is spontaneous and leads to the formation of co-continues phases, with final equilibrium compositions being nonetheless the same. The ratio of the equilibrium phases for each case (which can be determined by lever rule), are however different. For systems undergoing spinodal decomposition, provided the fraction of the minor phase is sufficient, polymer-rich and the polymer-poor phases that form would be initially completely interconnected. However, the coarsening process, as the equilibrium state is reached may alter this, unless structures that develop are locked in by solidification (for example by vitrification, or crystallization).

Figure 9b shows schematically the different morphologies that develop during phase separation by different paths described in Figure 9a, and the eventual, completely phase separated, equilibrium condition. The kinetics of the phase

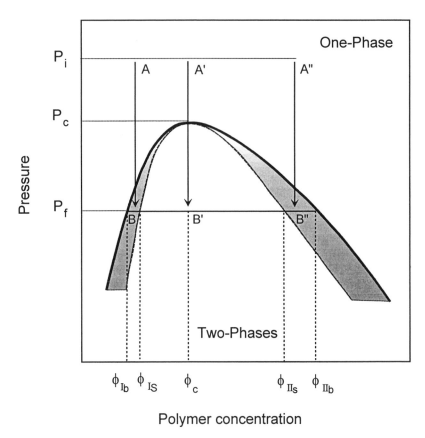

Figure 9a. Pressure-induced phase separation by different paths resulting in the same equilibrium phase compositions ϕ_{Ib} and ϕ_{IIb}. The solution is initially one-phase at P_i and enters the two-phase region upon reducing the pressure to P_f. Shaded are is the metastable region. New phase formation and growth proceeds by either nucleation and growth (paths AB, and A"B") or by spinodal decomposition (path A'B') mechanism.

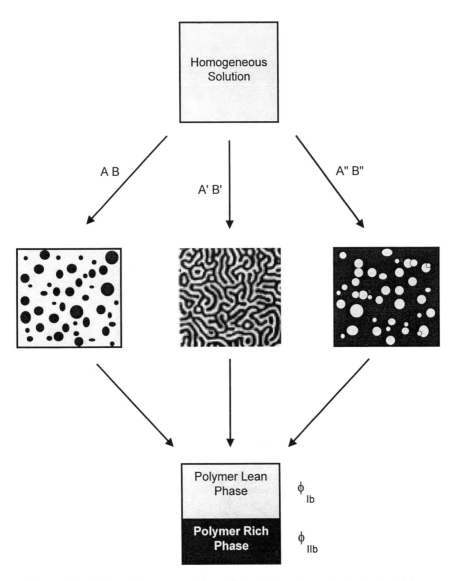

Figure 9b. Schematic representation of the different morphologies (particles, interconnected network, porous matrix) that are formed during phase separation, and the eventual two-phase state in equilibrium. Under certain circumstances, the kinetics of phase separation may prevent reaching the final equilibrium state, thereby leading to different morphologies.

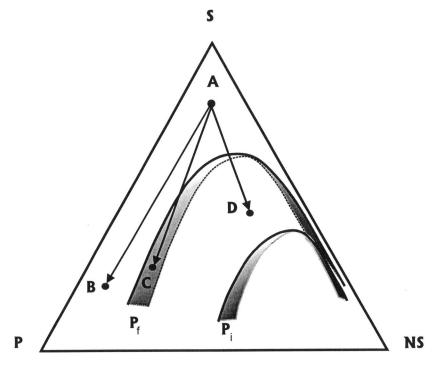

Figure 10. Pressure-induced phase separation in a ternary system of "polymer + solvent + non-solvent" [such as polyethylene + pentane + carbon dioxide (see Figure 7)]. Compositions corresponding to points A, B, C, and D are all in one-phase region at an initial high-pressure condition of P_i. But when pressure is reduced to P_f, the solutions with compositions C and D will enter the metastable and unstable regions, and will undergo phase separation by nucleation growth and spinodal decomposition, respectively.

separation processes however may prevent reaching the final equilibrium state and non-equilibrium states may be locked-in. It should be emphasized that depending upon the concentration of the solution, and the path followed, pressure-induced phase separation may be utilized to form polymers in powder form, or as porous materials with differing overall porosity, pore size and distributions offering a variety of new opportunities for processing of polymers from near and supercritical fluids to produce materials with desired morphologies. For ternary systems, how a point in the one-phase region may lie in the metastable or unstable regions upon a decrease in pressure can be easily visualized in Figure 10 which is a schematic representation of Figure 7 showing the phase boundaries at two different pressures. Points A, B, C, and D are all in the one phase region at high pressure P_i, but when the pressure is reduced to P_f, point C will in the metastable and D will be in the unstable regions.

As examples of different morphologies that can be produced, Figure 11a shows polyethylene crystals formed inside the microporous cavity (pithole) of a wood species impregnated by a 1 % solution of PE (16,400) followed by a rapid de-pressurization to precipitate the polymer within the wood's microporous network. Figure 11b shows an electron micrograph of a microporous material produced by pressure-induced phase separation from a natural (cellulosic) polymer derivative swollen in carbon dioxide + solvent system.

Kinetics of Pressure-induced Phase Separation

Kinetics of phase separation provides information on the time scale of new phase formation and growth which is important for rational process design and for achieving target material properties. Phase separation by nucleation and growth and by spinodal decomposition can lead to different morphologies. Even though for solutions at the critical polymer concentration, the phase separation proceeds by spinodal decomposition, at any other concentration, to enter the spinodal requires that the metastable regions are entered first where demixing starts with nucleation and growth (Figure 9). Entering the spinodal boundary, without extensive phase separation by the nucleation and growth process taking place, would require extremely rapid, deep-penetration into the region of immiscibility. For solutions away from the critical concentration, to enter the spinodal by temperature or solvent jump is not easy since extremely rapid changes in temperature may not be realized throughout the bulk of solution due to heat transfer limitations, and since the progress of solvent induced phase separation (i.e., addition of non-solvent to the system) will be limited by mass transfer. It is in this respect that pressure-induced phase separation differs and of great significance since pressure can be changed very rapidly, and the spinodal regime may be entered even at concentrations away from the critical concentration. This is demonstrated in Figure 12. The solutions following paths AB, A'B', and A"B", provided the metastable regions are passed extremely rapidly, will all undergo phase separation by the spinodal decomposition mechanism. The eventual equilibrium phase compositions will however, again, be the same for all these paths, namely ϕ_{Ib} and ϕ_{IIb}.

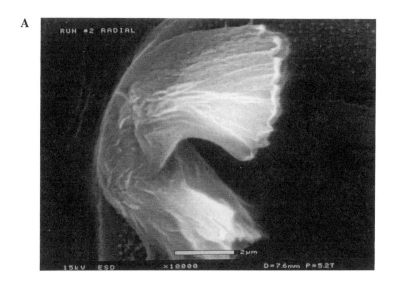

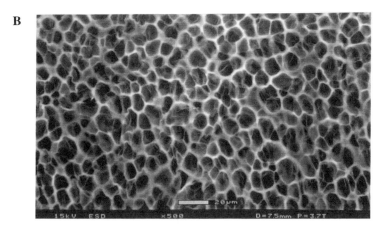

Figure 11. A (Top). Polymer (polyethylene) crystal formed by pressure-induced phase separation from a polymer solution of low concentration. **B (Bottom).** Porous polymer matrix (cellulose derivative) formed by pressure-induced phase separation from a solution at high polymer concentration.

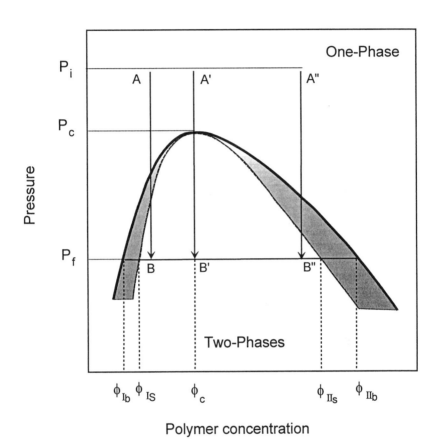

Figure 12. Pressure-induced phase separation by different paths. The solution is initially one-phase at P_i and enters the unstable two-phase region upon reducing the pressure extremely rapidly to P_f. The metastable region is passed very rapidly without much progress by nucleation and growth mechanism and new phase formation and growth proceeds essentially by the spinodal decomposition for all paths.

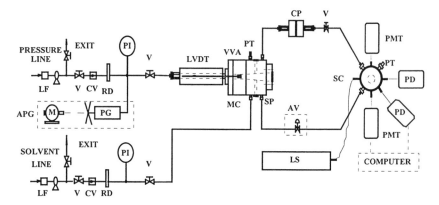

Figure 13. Experimental system to study kinetics of pressure-induced phase separation by time-resolved light scattering. [Legend: APG = automated pressure generator; AV = air-actuated valve; CP = magnetic recirculation pump; CV = check valve, LF = line filter; LS = laser source (Argon ion or He-Ne laser); LVDT = linear variable differential transformer; M = motor; MC = main cell; PD = photodetector; PG = pressure generator; PI = pressure indicator; PMT = photomultiplier tube; PT =pressure transducer; RD = rupture disc; SC = scattering cell; SP = sample port; V = valve; VVA variable-volume assembly]

In our research group, we have been interested in both slow and rapid pressure-drop experiments to follow the progress of phase separation. This is achieved by time-resolved light scattering using a unique experimental system shown in Figure 13. The system is a two-cell arrangement consisting of a variable-volume view-cell and a fiber-optic scattering-cell of much smaller volume. They are coupled with a magnetic recirculation loop and a fast opening air-actuated valve. Initially homogeneous one-phase solution is maintained in both cells with the help of the recirculation pump. Then, the two cells are isolated by closing the valves and the pressure in the variable volume cell is lowered by moving the piston to a new value. Then, the air actuated valve is opened to bring about a fast pressure drop. The progress of phase separation is monitored as a function of time by monitoring the transmitted and scattered light intensities from different angles over time intervals ranging from milliseconds, to seconds, or minutes by optical sensors positioned around the scattering cell.

Figure 14a shows the change in the 90 degree scattered light intensity over a 200 ms observation time for PE (M = 16,500) + pentane system, 5 % by mass polymer, undergoing a rapid pressure quench from 7 MPa to 5.4 MPa. After such a pressure drop experiment, the valves are opened, and the system is brought back to initial pressure and homogeneous one-phase conditions in both cells. The experiment is then repeated, but this time before the valve is opened, pressure in the view cell is reduced to an even lower value to accomplish deeper penetrations into the region of immiscibility. Such pressure-drop or pressure-jump experiments can be conducted repetitively without changing the cell content. Figure 14b shows the evolution of the new phase formation for the same system quenched to 4.7 MPa where scattered light intensity increase takes place in a much shorter time interval. Analysis of a series of such experiments provide insight as to the time scale of phase separation and time-scale for the growth of domains that would scatter an incident light of known wavelength (here 514 nm, the wave-length of the Argon-ion laser used). Figure 15 is a comparative plot covering a longer time interval (upto 1000 ms) and a range of pressure-quench experiments. The figure demonstrates that if penetration is not sufficient, new phase growth either does not take place or proceeds rather slowly, whereas with deep-quenches, phase growth becomes very rapid. In fact, if pressure quench is below a characteristic end-pressure, phase separation becomes very fast as reflected by the dramatic increase in the rate of change in scattered light intensity. The quench pressure where this transition takes place depends on the initial polymer concentration [30], and identifies a dynamic phase boundary (inside the binodal boundary) below which, within experimental limitations, the observable time-scale for phase separation is no longer altered by further penetration to lower pressures. It suggests that even in solutions at concentrations other than the critical polymer concentration, it may be possible to enter the spinodal by rapid quench to pressures below a characteristic pressure. In such a case, the pressure-reduction path passes through the metastable region rapidly before nucleation and growth can progress to any appreciable extend, and spontaneous phase separation region is entered.

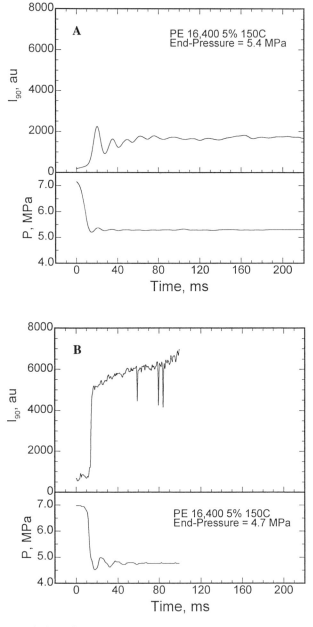

Figure 14. Variation of pressure and the 90 degree scattered light intensity with time during pressure-induced phase separation in PE (M = 16,500) + pentane solutions at 150 °C. Polymer concentration is 5 mass %. **A (Top)** Initial pressure = 7 MPa. Final (quench) pressure =5.4 MPa. **B (Bottom).** Initial pressure = 7 MPa. Final (quench) pressure = 4.7 MPa.

Extremely fast quench experiments with deep penetration into the region of immiscibility are rare in the literature. Recently, it has been proposed that in such experiments, polymer solutions may display a dynamic symmetry line and undergo a viscoelastic spinodal decomposition process [32, 33]. In very fast quench experiments, it is argued that the elastic nature of the polymers may delay phase separation. The notion is that there may not be enough time for the coils to agglomerate as they approach each other, rather they bounce back and the system may appear to be in solution while it should have undergone phase separation.

More recent effort in our laboratory is being devoted to the evaluations of the angular dependence of time-evolution of scattered light intensities which can provide additional information on the kinetics and the mechanism of phase separation [34]. In this respect, studies on scattering at low angles (i.e., less than 15 degrees) is particularly important. The well known Cahn-Hilliard theory of spinodal decomposition [35] which assumes that the dynamics of the phase separation process is governed by diffusion in a chemical potential gradient has been extensively applied to phase separation in polymer blends [36-40] and predicts that in the initial stages of phase separation, scattering intensity increases exponentially according to

$$I (q, t) = I (q, 0) \exp [2R(q)t] .$$

Here q is the wave number of growing fluctuations given by

$$q = 4\pi/\lambda \, \text{Sin} \, (\theta/2)$$

where λ is the wavelength and θ is the scattering angle, and the the rate of growth of concentration fluctuations $R(q)$ is given by

$$R (q) = D^* \, q^2 \, [1-q^2/2q_m^2]$$

where D^* is the apparent diffusivity and q_m is the wave number corresponding to maximum growth rate of fluctuations. The value of q which makes $R(q)$ maximum is the most probable wave number of fluctuations with the highest rate of growth. The linearized Cahn-Hilliard theory predicts a time-independent q_m . The effective diffusivity can be calculated from plots of $R(q)/q^2$ versus q^2 as the limiting value of $R(q)/q^2$ as q approaches to zero, i.e.,

$$D^* = \lim_{q \to 0} \{R(q)/q^2\}$$

and the spinodal is then determines as the condition where D^* becomes zero. By conducting pressure-induced phase separation with varying quench depths, verification that below a certain pressure, scattered light intensities go through a maximum as a function of wave vector would provide the theoretically expected verification for spinodal mechanism for phase separation.

From rapid deep-quench experiments in which spinodal decomposition is entered, provided the linear theory is applicable, monitoring the time evolution of scattered light intensities at a fixed angle can also be used to assess the spinodal pressure. At a fixed angle, for a given incident light, q becomes fixed, and R becomes proportional to D*. For a given end-pressure corresponding to a quench, from the linear portion of the ln I (t) versus t curves, R can be evaluated from the slopes. These calculations can be repeated for additional quenches with different end-pressures. From an extrapolation of a plot of R versus the end-pressure, the pressure where R (or D*) becomes zero can be identified as the spinodal pressure. This is demonstrated in Figure 16 for polyethylene (M=108,000) solutions in pentane for two concentrations. These experiments can then be repeated for all concentrations, and the spinodal envelope may be generated. Figure 17 demonstrates the spinodal boundary for polyethylene (M=108,000) + pentane solutions generated from extrapolations of the light scattering data at 60 degrees, and compares it with the binodal (cloud point curve). Theoretically, the spinodal should merge with the binodal. This is however not observed in Figure 17. Possible reasons for this are the fast nature of the quenches that are imposed which lead to a viscoelastic spinodal decomposition, or simply the non-linear nature or the progress of the spinodal decomposition under rapid quench experiments and thus the inapplicability of the conventional Cahn-Hilliard analysis. Further experimentation is needed and currently is underway [34] in our laboratory for more extensive documentation of the details and the consequences of the kinetics of phase growth associated with pressure-induced phase separation in "polymer + supercritical fluid", and "polymer + solvent + supercritical fluid" systems. Indeed, low-angle light scattering results in polystyrene + methylcyclohexane solutions subjected to rapid pressure quenches undergo non-linear spinodal decomposition [34], the wave vector corresponding the the scattering maximum is not time indpendent and moves to lower q values with progress of phase separation. Polymer solutions are dynamically assymetric, and unlike polymer blends, may in general display non-linear spinodal decomposition.

Concluding Comments

Supercritical fluid processing in the polymer industries is indeed of great interest for a diversity of applications ranging from synthesis, to property modifications and processing. The present paper has focused primarily on miscibility and phase separation, which are the common crucial steps in any process involving these fluids. Our understanding of the diverse ways in which binary fluid mixtures can be used to bring about miscibility and phase-separation, and our improved understanding of the kinetic mechanism(s) of the phase separation processes are opening up many possibilities for practical applications. Binary fluid mixtures can be used to affect selectivity towards a polymer, or introduce a greater sensitivity of the system to changes in external parameters such as temperature or pressure. Morphological control of materials that are produced by pressure-induced phase separation is

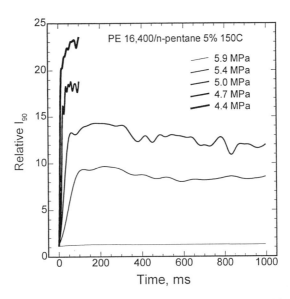

Figure 15. Variation of the 90 degree scattered light intensity with time during pressure-induced phase separation in PE (M = 16,500) + pentane solutions at 150 °C. Polymer concentration is 5 mass %. Initial pressure = 7 MPa. Final (quench) pressures are as indicated.

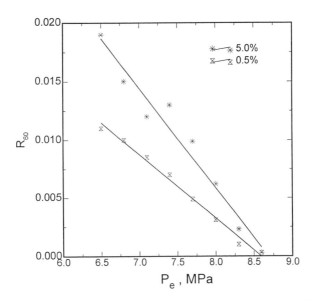

Figure 16. Analysis of the scattering data at a fixed angle for polyethylene (M =108,000) solutions in n-pentane undergoing pressure-induced phase separation at 150 °C . According to the linear Cahn-Hilliard formalism, the limit of R = 0 defines the spinodal pressures (see text).

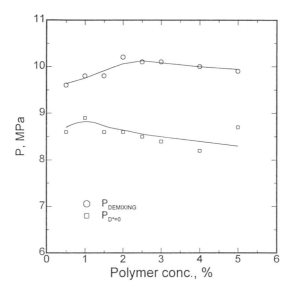

Figure 17. Comparison of the binodal and the spinodal pressures for Polyethylene (M =108,000) + pentane system at 150 °C.

intimately linked to the time scale of phase separation and its relationship with vitrification and / or crystallization processes. Pressure-induced phase-separation is a powerful technique to bring about very fast quenches from polymer solutions to lock in target structures and therefore has great implications for manufacture of specialty materials.

Acknowledgments

This chapter is based on research activities that has been supported by Industry, NSF, USDA, and the Supercritical Fluids Program (SFEST) at the University of Maine.

Literature Cited

1. Kiran, E.; Levelt Sengers, J. M. H., Eds., *Supercritical Fluids. Fundamentals for Application*, Kluwer Academic Publishers: Dordrecht, **1994**.
2. McHugh, M. A.; Krukonis, V. *Supercritical Fluid Extraction*, 2nd Ed., Butterworth: Boston, **1994**.
3. *Extraction of Natural Products Using Near-Critical Solvents*, King, M. B.; Bott, T. R., Eds., Chapman & Hall: New York, **1993**.
4. Anon. Modern Plastics, **1995,** 72 (1), 63.
5. Kiran, E. in *Supercritical Fluids. Fundamentals for Application*, Kiran, E.; Levelt Sengers, J. M. H., Eds., Kluwer Academic Publishers: Dordrecht, **1994**; pp. 541-588.

6. Xiong, Y.; Kiran, E. *Polymer*, **1995**, 36 (25), 4817-4826.
7. Kiran, E.; Gökmenoglu, Z. *J. Appl. Polm. Sci.*, **1995**, 58, 2307-2324.
8. Xiong, Y.; Kiran, E. *Polymer*, **1994**, 35(20), 4408-4415.
9. Xiong, Y.; Kiran, E. *J. Appl. Polm. Sci.*, **1995**, 55, 1805-1818.
10. Kiran, E.; Balkan, H. *J. Supercrit. Fluids*, **1994**, 7, 75-86.
11. Xiong, Y.; Kiran, E. *J. Appl. Poly. Sci.*, **1994**, 53, 1179-1190.
12. Kiran, E.; Xiong, Y.; Zhuang, W. *J. Supercrit. Fluids*, **1994**, 7, 283-287.
13. Kiran, E.; Zhuang, W. *J. Supercrit. Fluids*, **1994**, 7, 1-8.
14. Kiran, E. In *Innovations in Supercritical Fluids -Science and Technology*, K. Hutchenson, and N. Foster, Eds, ACS Symposium Series No. 608; American Chemical Society, Washington DC, **1995**, pp 380-401.
15. Kiran, E.; Malki, K.; Pöhler, H. *Proc. Amer. Chem. Soc. Div. Polymeric Materials Sci. & Eng.* **1996**, 74, 231-232.
16. Kiran, E.; Gökmenoglu, Z. *Proc. Amer. Chem. Soc. Div. Polymeric Materials Sci. & Eng.* 1996, 74, 406-407.
17. Doak, K. W. in Concise Encyclopedia of Polymer Science & Engineering, John Wiley and Sons,: New York, **1990**; p. 350.
18. DeSimone, J. M. ; Maury, E. E.; Menceloglu, Y. Z.; McClain, J. B.; Romack, T. R. *Science*, **1994**, 265, 356.
19. Romack, T.J.; DeSimone, J. M. *Proc. Amer. Chem. Soc. Div. Polymeric Materials Sci. & Eng.* **1996**, 74, 428-429.
20. Costello, C. A.; Berluche, E.; Han, S. J.; Sysyn, D. A.; Super, M. S. ; Beckman, E. *Proc. Amer. Chem. Soc. Div. Polymeric Materials Sci. & Eng.* 1996, 74, 430-431.
21. Chang, E. L. *Physica*, **1986**, 139 &140B, 885-889.
22. Freunfelder, H., et.al., *J. Phys. Chem.* **1990**, 94(3), 1024-1037.
23. Eugster, H. P. *Amer. Mineral.* **1986**, 71, 655.
24. Walther, J. V. *Pure & Appl. Chem.* **1986**, 58(12), 1598.
25. *Handbook of Chemistry and Physics*, 71st ed., Life, D. R. Ed.; CRC Press: Baco Raton, **1990**; pp. 6/48-6/69.
26. Kiran, E. Paper presented at *Symposium on Thermosetting Polymers*, Princeton University, September 22-23, **1995**.
27. Rader, P.; Stockel, R. F. in *Plastics, Rubber and Paper Recycling. A Pragmatic Approach.*, Rader, C. P. et al. Eds,, *ACS Symposium Series 609*, American Chemical Society: Washington, DC, **1995**; pp. 2-10.
28. Babinchak, S. R.; in *Proc. Plastic Waste Management-Recycling Alternatives*, *SPE RETEC*, White Havens, PA, October 17-18, **1990**; paper 15.
29. Anon. *Chem. Eng. Prog.* November **1991**, pp.22-23.
30. Zhuang, W. *Ph.D. Thesis*, University of Maine, 1995.
31. Pöhler, H.; Kiran, E. to be published.
32. Tanaka, H. *J. Chem. Phys.* **1994**, 100(7), 5323-5337,
33. Tanaka, *Int. J. Thermopysics*, **1995**, 16(2), 371-380.
34. Xiong, Y. *Ph.D. Thesis*, University of Maine, in progress.
35. Cahn, J. W., Trans. Mettall. Soc. AIME., **1968**, 242, 166
36. Snyder, H. L.; Meakin, P. *J. Polym. Sci. Polymer Symp.* **1985**, 73, 217-239.

37. Hashimoto, *Phase Transitions,* **1988**, 12, 47-119.
38. Inoue, T.; Ougizawa, T. *J. Macromol. Sci.-Chem.* **1989**, A26(1), 147-173.
39. L. A. Utracki, in Interpenetrating Polymer Networks, D. Klempner; Sperling, L.H., Eds.; *ACS Adv. Chem. Ser. 239,* American Chemical Society: Washington, DC, **1994**; pp. 77-123.
40. Perreault, F.; Prud'homme, R. E. *Polym. Eng. &Sci.,* **1995**, 35(1), 34-40.
41. Cheng, H.; Fernandez, M. E.P.; Zolwegg, J. A.; Street, W. B. *J. Chem. Eng. Data,* **1989**, 34, 319-323.
42. Kiran, E.; Pöhler, H.; Xiong, Y. *J. Chem. Eng. Data.* **1996**, 41, 158-165
43. Kiran, E.; Zhuang, W.; Sen, Y.L. *J. Appl. Polm. Sci.* **1993**, 47, 895.
44. Kiran, E.; Pöhler, H. to be published

Chapter 2

Estimation of Solid Solubilities in Supercritical Carbon Dioxide from Solute Solvatochromic Parameters

David Bush and Charles A. Eckert[1]

School of Chemical Engineering and Specialty Separations Center,
Georgia Institute of Technology, Atlanta, GA 30332-0100

Good estimating methods for solid solubilities in supercritical fluids are valuable tools in supercritical fluid processing. Experimental measurements of solid solubilities are time-consuming and may have large uncertainties. A linear solvation energy relationship has been developed for the enhancement factor, the ratio of solubility of a solid in an SCF to its solubility in an ideal gas. The solute parameters used are π_2^*, the dipolarity/polarizability, $\Sigma\alpha_2^H$, the effective hydrogen-bond acidity, $\Sigma\beta_2^H$, the effective hydrogen-bond basicity, and V_x, the McGowen characteristic volume. A database of 37 solids containing sublimation pressure, solubility in carbon dioxide, and solute solvatochromic parameters was assembled and regressed. The average deviation of the estimated solubilities is 55 percent, which in some cases appears within the experimental uncertainty of the measurements. With over 1000 experimentally determined solute solvatochromic parameters available, this equation greatly expands the small database from which this correlation was developed. In addition, this empirical equation gives some physical insight to the nature of carbon dioxide - solute interactions.

Carbon dioxide is a benign and inexpensive solvent that is used commercially for the decaffeination of coffee beans and extraction of flavors from hops. As a supercritical fluid (SCF), the solvent strength of CO_2 can easily be adjusted with small changes in pressure and without phase transitions. Because of these properties, CO_2 is a useful processing solvent for extractions and reactions because it can be easily separated from the desired product (*1*). However, large scale use of carbon dioxide has not yet gained wide acceptance, and only a few commercial

[1]Corresponding author

processes are operational. Part of the reason is the lack of phase equilibria data or a suitable predictive model needed for feasibility studies (2). The solubility of more than 160 pure solids have been measured in supercritical CO_2 at various temperature and pressures. The compounds range from model compounds with small substituent changes to natural products of peculiar molecular structure. This paper will examine some of the techniques used to obtain equilibrium data, show how these data are modeled, and present a novel technique for correlating these data to obtain predictions.

Experimental Data

Of the many techniques presented in the literature, most give similar results when the solubility is high, *i.e.* when the solubility is greater than 10^{-2} mole fraction. A comparison of the solubility of naphthalene shows that the experimental data from different laboratories agree within 10 percent (2,3). However, when the solubility is low, the differences between published results can be quite large. For example, the reported values for octacosane solubility in CO_2 at 35°C differ by more than a factor of 10. This is one of the few systems where data from many different laboratories can be compared. These results are shown in Figure 1 along with error bars of 15 percent, an uncertainty greater than that suggested by the investigators. It is not clear why the data are so disparate; these measurements are not trivial. The majority of all CO_2-solid equilibria measurements are investigated by variations on transpiration or spectroscopic techniques, each of which have their own limitations.

The continuous flow (CF) or transpiration method for measuring solubilities appears the most reliable and consistent. In this technique the SCF is passed through a packed bed of solid slowly enough so that equilibrium is reached. The challenge is to measure the amount of SCF and solid in the stream leaving the bed.

The amount of SCF can be measured either before or after the packed bed. By pumping the SCF with an accurate metering pump, the amount of SCF can be calculated from the displaced volume and the temperature and pressure at the pump head (4,5). This technique is dependent on the accuracy of the equation of state used for the particular SCF. When the SCF is a mixture, experimental density measurements should be used. A more accurate approach is to measure the SCF leaving the apparatus with a wet gas flow meter (6).

In order to determine the amount of solid dissolved in the SCF, the saturated solution is usually passed through a micrometering valve, which reduces the pressure to atmospheric. The solute, no longer soluble in the gas, drops out of solution and into a trap placed after the micrometering valve. After 30 - 100 mg of solute has been collected, the trap is weighed. Numerous examples of this technique exist in the literature (7-9). It is a time-consuming and difficult experiment. One solubility measurement may require 4 to 6 hours for a sufficient amount of solute to collect.   Also, the micrometering valve requires constant adjustment to prevent the solute from plugging in the needle.   Heating the valve to above the melting temperature of the solute helps, but it does not always alleviate the problem.   In addition, frequent plugging can make it difficult to determine accurately the amount of SCF passed.

The uncertainties with this approach increase as the solubility decreases.   In general, uncertainties can be as little as 5 percent for mole fractions greater than 10^{-2} and as much or greater than 50 percent for mole fractions less than 10^{-4}.

Spectroscopic measurement provide an alternative technique that allows for measuring small solubilities *in situ*.   The concentration of the solute, S, can be determined from Beer's law

$$S = \frac{A}{\varepsilon l} \tag{1}$$

where A is the measured absorbance at some wavelength, l is the path length, and ε is the molar absorptivity of the solute.   Solubilities have been determined using IR and UV-vis transmissions spectroscopy (*10,11*).   Measuring solubilities *in situ* has an added advantage of watching the diffusion of the solute to the bulk phase in real time, ensuring that equilibrium is obtained.   The drawback to this method is the extensive calibration required for the molar absorptivity.   The molar absorptivity is dependent on the density of the solvent and can be substantially different from that in liquids.   Solubility measurements not accounting for this change could   have errors as high as 300 percent (*12*).

Calculation of Solid-SCF Equilibria

Two alternative approaches to modeling solid-SCF equilibria are to treat the supercritical fluid solution as either a compressed gas (CG) or as an expanded liquid (EL) solution. The solid is generally modeled as a pure incompressible phase.   At equilibrium, the pure solid and the solid in solution have equal fugacities.   The derivation for fraction dissolved, y_2, from both models has been given by Mackay and Paulaitis (*13*).   The CG model is

$$y_2 = \frac{P_2^S \exp \dfrac{v_2^S P}{RT}}{\phi_2 P} \tag{2}$$

where P_2^S is the vapor pressure of the solid, v_2^S is the molar volume of the solid, ϕ_2 is the fugacity coefficient of the solid in the SCF phase. This equation holds if v_2^S is constant and P_2^S is small.   This model has three variables: Two are physical properties of the solid and one is a mixture property. The molar volumes are available for most solids, or they can be accurately estimated. The vapor pressures are available for many compounds whose vapor pressure is above 10 mPa. The fugacity coefficient is a function of the P-ρ-T properties of the solution, which are rarely available, and can be predicted only approximately from an equation of state. For example, the Peng-Robinson equation of state with van der Waals one-fluid mixing rules predicts the solubility of anthracene in CO_2 at 35°C and 172 bar to be 7.3 x 10^{-4}. This is a substantial difference from the measured value of 8 x 10^{-5} (*14*). More often, one or two adjustable parameters in the equation of state are regressed to the experimental data. To match the data of Kosal and Holder, the interaction parameter k_{ij} must be set to 0.075.

Several researchers have developed density-dependent mixing rules for asymmetric systems. With these models, it has been shown that the number of parameters can be reduced and still achieve good results (*15-17*). The data sets used were quite limited, mostly polynuclear aromatics with few functionalities.

The EL model is

$$y_2 = \frac{f_2^{0S} \exp \dfrac{v_2^S \left(P - P^0 \right)}{RT}}{\gamma_2 f_2^{0L} \exp \displaystyle\int_{P^0}^{P} \dfrac{\bar{v}_2 \, dP}{RT}} \qquad (3)$$

where f_2^0 is the pure component fugacity, P^0 is the pressure of the standard state reference, γ_2 is the activity coefficient at P^0, and $\bar{v}_2$ is the partial molar volume of the solid in the SCF. The pure component properties for this model are more accessible because the ratio of the pure solid fugacity to the pure subcooled liquid fugacity can be approximated by

$$\frac{f_2^{0S}}{f_2^{0L}} = \exp\left[\frac{\Delta H_2^{fus}}{R} \left[\frac{1}{T_2^{fus}} - \frac{1}{T} \right] \right] \qquad (4)$$

where ΔH_2^{fus} is the heat of fusion of the solid at its melting temperature, T_2^{fus}. The EL model also requires two mixture properties, γ_2 and $\bar{v}_2$. As yet, there are no accurate methods for determining activity coefficients in carbon dioxide. Ziger and Eckert used regular solution theory with 2 adjustable parameters to regress solubility data of nonpolar compounds in CO_2, C_2H_6, and C_2H_4 (*18*). The other problem with the EL model lies in calculating the exponential in the denominator of equation (3). The partial molar volume in the highly compressible region, near the critical point, is large, negative, and highly density dependent, which can make the exponential of the integral as large as a factor of 5. A few partial molar volumes have been determined from density measurements (*19,20*) and SCF chromatography (*21,22*) but have not been correlated.

Gurdial *et al.* (*23*) examined some empirical and semi-empirical models that correlated the enhancement factor, *E*, in supercritical fluids. The enhancement factor, also the ratio of the solubility in a SCF to the solubility in an ideal gas, is defined as

$$E = \frac{yP}{P^{sub}} \qquad (5)$$

Gurdial *et al.* showed that for nonpolar compounds, semilog plots of E versus density could be approximated by straight lines that were parallel and equally spaced at different temperatures. However, not all compounds showed this dependence, and polar compounds showed more curvature and no true distinct parallel lines. By including the vapor pressure of the solid, the enhancement factor accounts for the solute-sovent interactions, which, for water, have been modeled successfully with linear solvation energy relationships (*24*). In Figure 2, the solubility is plotted versus vapor pressure for several solids. At the same vapor pressure, there can be significant differences in solubility for two solids which can be attributed to solid-carbon dioxide interactions. For the solids with a solubility greater than 10^{-2}, there are also solute-solute interactions, but the solute-solvent interactions dominate.

Solvatochromic Parameters for Linear Solvation Energy Relationships (LSER)

The model for using solvatochromic parameters to describe a molecule of an ideal gas being solvated has been described by Carr (*25*). The two steps involved are (1) the cavity formation process and (2) the solute-solvent interactions. The mathematical model can be written as

$$XYZ = XYZ_0 + \text{cavity formation energy} + \sum \text{solute-solvent interactions} \qquad (6)$$

The function *XYZ* is some configurational property, such as a partition coefficient. The cavity formation energy is usually described by the cohesive energy density, if the LSER is for correlating solvents, or the molar volume, if the LSER is correlating solutes. The solute-solvent interactions are usually described by scales of dipolarity, polarizability, and hydrogen bonding. The intercept XYZ_0 accounts for everything else, including the fact that the scales are arbitrarily scaled.

Kamlet, Taft, and their coworkers established a set of solvent scales by observing the solvatochromic shifts of nitroaromatic probes in various solvents (*26*).

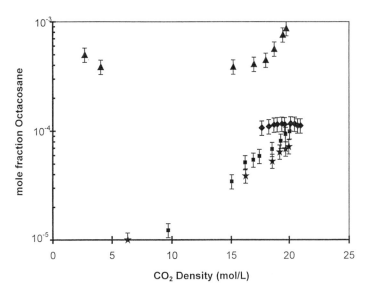

Figure 1. Solubility Measurements of Octacosane in CO_2 at 35°C. Data from ◆
(36), ■ (37), ▲ (38), ★ (39).

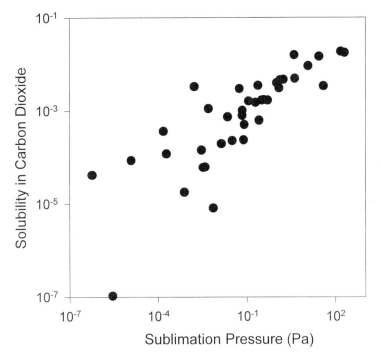

Figure 2. Sublimation Pressure Influence on Solubility at 35°C and 150 bar.

They were able to characterize solvents by the dipolarity/polarizability, π^*, hydrogen bonding acidity, α_{KT}, and hydrogen bonding basicity, β_{KT}. These scales are orthogonal and are used to characterize the solute-solvent interactions. Along with the Hildebrand solubility parameter, a term used to account for the cavity formation, these scales proved useful for characterizing free energy processes of different solvents. For example, the free energy of solution of nitromethane in 18 nonprotonic solvents is given by

$$\Delta G_s = \left(-1.54 + 0.21\delta_H - 0.56\pi^* \right) \text{kcal} / \text{mol} \qquad (7)$$

Since neither nitromethane nor the solvents studied can donate a proton for a hydrogen bond, the coefficients for α_{KT} and β_{KT} are set to zero.

Kamlet, Taft, and coworkers applied the solvatochromic parameters, determined from bulk properties, to solutes, and they were able to correlate retention times for several solutes in reversed-phase liquid chromatography. Subsequent work by Carr, Abraham, and coworkers showed that a separate set of solute parameters was necessary for solutes that self-associate in bulk (*25,27*). This lead Carr and coworkers to develop techniques for using chromatography to rapidly measure solute "solvatochromic" scales (*28*).

Development of the model

A database of enhancement factors of 37 solids in CO$_2$, limited by the availability of sublimation pressures, was collected from the literature and are shown in Table I. The solute parameters of Abraham *et al.* for the 37 solids are listed in Table II (*29-31*). The parameters are V_x, McGowen's intrinsic volume for the solute; $\Sigma\alpha_2^H$, the effective hydrogen bond acidity; $\Sigma\beta_2^H$, the effective hydrogen bond basicity; π_2^*, the dipolarity/polarizability; and R_2, the excess index of refraction. The subscript 2 identifies these parameters as solute-specific. It was found, using multiple linear regression, that the model

$$\log E = s\pi_2^* + rR_2 + a\Sigma\alpha_2^H + b\Sigma\beta_2^H + vV_x + E_0 \qquad (8)$$

was appropriate for a given temperature and pressure. The coefficients were temperature and pressure dependent, as expected, since they describe the part carbon dioxide plays in the solvation process. The intercept, E_0, is also an adjustable parameter dependent on temperature and density. It has been shown that the solvatochromic parameters for supercritical CO$_2$ are dependent on temperature and density (*32*).

Because the Kamlet-Taft scale and the Abraham scale are not directly compatible, a linear temperature, pressure, and density dependence was chosen arbitrarily so that the coefficients in equation (8) become

Table I. Solute Parameters

Compound	R_2	π_2^H	$\Sigma\alpha_2^H$	$\Sigma\beta_2^H$	V_x
acridine	2.356	1.33	0.00	0.58	1.414
anthracene	2.290	1.34	0.00	0.26	1.454
benzoic acid	0.730	0.90	0.59	0.40	0.932
biphenyl	1.360	0.99	0.00	0.22	1.324
caffeine	1.500	1.60	0.00	1.35	1.363
carbazole	1.787	1.50	0.35	0.24	1.315
carbon tetrabromide	1.190	0.94	0.00	0.00	0.950
chrysene	3.027	1.73	0.00	0.33	1.823
dibenzofuran	1.407	1.02	0.00	0.17	1.274
dibenzothiophene	1.959	1.31	0.00	0.18	1.379
dimethylnaphthalene, 2,3-	1.431	0.95	0.00	0.20	1.367
dimethylnaphthalene, 2,6-	1.329	0.91	0.00	0.20	1.367
dimethylnaphthalene, 2,7-	1.329	0.91	0.00	0.20	1.367
eicosanoic acid	0.005	0.60	0.60	0.45	3.001
fluoranthrene	2.377	1.55	0.00	0.20	1.585
fluorene	1.588	1.06	0.00	0.20	1.357
fluorenone, 9-	1.370	1.75	0.00	0.49	1.587
hexachloroethane	0.680	0.22	0.00	0.06	1.125
hexamethylbenzene	0.950	0.72	0.00	0.21	1.562
hydroquinone	1.000	1.00	1.16	0.60	0.834
hydroxybenzoic acid, o-	0.890	0.70	0.72	0.41	0.990
indole	1.200	1.12	0.44	0.22	0.946
methylbenzoic acid, m-	0.730	0.90	0.59	0.38	1.073
myristic acid	0.060	0.60	0.60	0.45	2.156
naphthalene	1.340	0.92	0.00	0.20	1.085
naphthaquinone, 1,4-	1.080	1.13	0.00	0.62	1.159
naphthol, 1-	1.520	1.05	0.61	0.37	1.144
naphthol, 2-	1.520	1.08	0.61	0.40	1.144
methylbenzoic acid, o-	0.730	0.90	0.60	0.34	1.073
methylbenzoic acid, p-	0.730	0.90	0.60	0.38	1.073
palmitic acid	0.035	0.60	0.60	0.45	2.437
phenanthrene	2.055	1.29	0.00	0.26	1.454
phenazine	1.970	1.53	0.00	0.59	1.372
pyrene	2.808	1.71	0.00	0.29	1.585
stearic acid	0.015	0.60	0.60	0.45	2.719
triphenylmethane	1.800	1.70	0.00	0.45	2.073
xylenol, 2,5-	0.840	0.79	0.54	0.37	1.057

Table II. Enhancement factor references

Compound	Vapor Pressure	Solubility
acridine	(40)	(41)
anthracene	(42-45)	(4,14,46-48)
benzoic acid	(49)	(50)
biphenyl	(42,51-53)	(54,55)
caffeine	(56)	(57,58)
carbazole	(59)	(48,60)
carbon tetrabromide	(61)	(41)
chrysene	(62)	(63)
dibenzofuran	(43,64)	(41,65)
dibenzothiophene	(43,66)	(65,67)
2,3-dimethylnaphthalene	(51,68)	(6)
2,6-dimethylnaphthalene	(51,68)	(6,69)
2,7-dimethylnaphthalene	(68)	(69)
eicosanoic acid	(70)	(71)
fluoranthrene	(72)	(63)
fluorene	(64,68,73)	(46,74)
9-fluorenone	(43)	(41)
hexachloroethane	(75)	(6)
hexamethylbenzene	(76)	(46,77)
hydroquinone	(78)	(47,79)
o-hydroxybenzoic acid	(80,81)	(8,82)
indole	(75)	(83)
m-methylbenzoic acid	(84)	(85)
o-methylbenzoic acid	(84)	(85)
p-methylbenzoic acid	(84)	(85)
myristic acid	(70)	(86,87)
naphthalene	(44,51,64,73,88)	(3,6,54,89)
1,4-naphthaquinone	(78)	(50)
1-naphthol	(90)	(47,91)
2-naphthol	(90)	(50,91)
palmitic acid	(70)	(86,87)
phenanthrene	(42,44,64)	(46)
phenazine	(40)	(34)
pyrene	(42,73,92)	(46)
stearic acid	(70)	(86,87)
triphenylmethane	(43)	(46)
2,5-xylenol	(93)	(94)

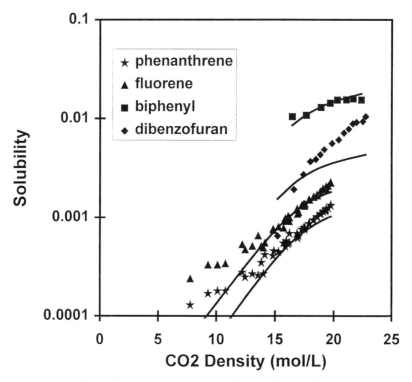

Figure 3. Correlation of Solubility in CO_2 at 35°C.

$$\beta = \beta_0 + \beta_1 T + \beta_2 \rho + \beta_3 P \qquad (9)$$

where β is a coefficient, T is the system temperature, P is the system pressure, and ρ is the density of pure CO_2 calculated from a modified BWR equation of state (*33*). The model has now increased in complexity to 21 adjustable parameters, many of which will not be significant. The multiple linear regression techniques of *best subsets* and *forward search* were used to determine the parameters that belong in the model. In *best subsets*, every possible model with 5 to 21 parameters was evaluated in terms of adjusted R^2, the variance of the data explained by the model adjusted for the number of parameters. With 10 parameters, the adjusted R^2 for the model peaked at 0.961. In *forward search*, variables are added to the model one at a time and only if they are significant. The final equation to predict enhancement factors in supercritical CO_2 is

$$\log E = E_0 + v_0 V_x + v_1 T_R V_x + d\rho_R$$
$$+ s_2 \rho_R \pi_2^* + r_2 \rho_R R_2 + a_1 T_R \Sigma \alpha_2^H + a_2 \rho_R \Sigma \alpha_2^H + b_2 \rho_R \Sigma \beta_2^H \qquad (10)$$

The coefficients with confidence limits are given in Table III. Note that the model now contains 9 parameters. The size and relative magnitude of the coefficients are expected. Van Alsten and Eckert have shown that v_0 should be positive (*34*). The signs of v_1 and d agree with the expression from Schmitt and Reid (*35*). The magnitudes of s, r, a, and b are much smaller but still significant, indicating weak localized interactions between CO_2 and the solute.

Table III. Model Parameters

Coefficient	Variable	Value	95 % Confidence
E_0		-1.04	±0.07
v_0	V_x	5.14	±0.16
v_1	$T_R V_x$	-2.88	±0.16
d	ρ_R	1.16	±0.05
s_2	$\rho_R \pi_2^{*H}$	0.52	±0.07
r_2	$\rho_R R_2$	-0.11	±0.04
a_1	$T_R \Sigma \alpha_2^H$	0.83	±0.16
a_2	$\rho_R \Sigma \alpha_2 H$	-0.22	±0.11
b_2	$\rho_R \Sigma \beta_2^H$	0.70	±0.10

Equation (10) gives an average error of 4.4 percent in log E, or an error of 6.8 percent in log y. The average error in y is 56 percent, but the error distribution is not normal. Figure 3 shows data for four of the thirty-seven solids and how well they correlate with the model. For most solids, the fit is good with the largest differences occurring for measurements in the low density region of CO_2 where the measurement error is also the largest. This model should give reasonable predictions for compounds whose the solute parameters are known and within the range of those fitted. For dibenzofuran and a few other compounds, the model gives a systematic deviation over the entire density range, indicating that a more complicated density dependence could better reproduce that data. Also, the measurements of most points are between 100 and 300 bar and 35 and 75°C. Extrapolation should be used with great caution.

Conclusion

The model affords a rough estimate of solubility provided that the sublimation pressure is known. The coefficients of the model make physical sense in that the size and density are the most important variables that affect the enhancement factor. The solute-solvent interactions make up a small, but still important contribution. This model does not yet extend to supercritical fluids with cosolvents, but it does provide a framework within which to proceed.

Acknowledgment

The authors gratefully acknowledge funding support from E.I. DuPont de Nemours Co.

References

1.   Brennecke, J. F.; Eckert, C. A. *AIChE. J.* **1989**, *35*, 1409.
2.   Hutchenson, K. W.; Foster, N. R. *Innovations in Supercritical Fluid Science and Technology*; Hutchenson, K. W.; Foster, N. R., Ed.; ACS: San Francisco, 1995; Vol. 608.
3.   Tsekhanskaya, Y. V.; Iomtev, M. B.; Mushkina, E. V. *Russ. J. Phys. Chem.* **1964**, *38*, 1173.
4.   Miller, D. J.; Hawthorne, S. B. *Anal. Chem.* **1995**, *67*, 273.
5.   Cowey, C. M.; Bartle, K. D.; Burford, M. D.; Clifford, A. A.; Zhu, S.; Smart, N. G.; Tinker, N. D. *J. Chem. Eng. Data* **1995**, *40*, 1217.
6.   Kurnik, R. T.; Holla, S. J.; Reid, R. C. *J. Chem. Eng. Data* **1981**, *26*, 47.
7.   Johnston, K. P.; Eckert, C. A. *AIChE J* **1981**, *27*, 773.
8.   Gurdial, G. S.; Foster, N. R. *Ind. Eng. Chem. Res.* **1991**, *30*, 575.
9.   Yun, S. L. J.; Liong, K. K.; Gurdial, G. S.; Foster, N. R. *Ind. Eng. Chem. Res.* **1991**, *30*, 2476.
10.   Ebeling, H.; Franck, E. U. *Ber. Bunsen Ges. Phys. Chem.* **1984**, *88*, 862.
11.   Zerda, T. W.; Wiegand, B.; Jonas, J. *J. Chem. Eng. Data* **1986**, *31*, 274.
12.   Rice, J. K.; Niemeyer, E. D.; Bright, F. V. *Anal. Chem.* **1995**, *67*, 4354.
13.   Mackay, M. E.; Paulaitis, M. E. *Ind. Eng. Chem. Fundam.* **1979**, *18*, 149.
14.   Kosal, E.; Holder, G. D. *J. Chem. Eng. Data* **1987**, *32*, 148.
15.   Chen, P.-C.; Tang, M.; Chen, Y.-P. *Ind. Eng. Chem. Res.* **1995**, *34*, 332.
16.   Sheng, Y.-J.; Chen, P.-C.; Chen, Y.-P.; Wong, D. S. H. *Ind. Eng. Chem. Res.* **1992**, *31*, 967.
17.   Rao, V. S. G.; Mukhopadhyay, M. *J. Supercrit. Fluids* **1990**, *3*, 66.
18.   Ziger, D. H.; Eckert, C. A. *Ind. Eng. Chem. Process Des. Dev.* **1983**, *22*, 582.

19.   Eckert, C. A.; Ziger, D. H.; Johnston, K. P.; Ellison, T. K. *Fluid Phase Equilib.* **1983**, *14*, 167.
20.   Eckert, C. A.; Ziger, D. H.; Johnston, K. P.; Kim, S. *J. Phys. Chem.* **1986**, *90*, 2738.
21.   Shim, J. J.; Johnston, K. P. *J. Phys. Chem* **1991**, *95*, 353.
22.   Gönenç, Z. S.; Akman, U.; Sunol, A. K. *J. Chem. Eng. Data* **1995**, *40*, 799.
23.   Gurdial, G. S.; Wells, P. A.; Foster, N. R.; Chaplin, R. P. *J. Supercrit. Fluids* **1989**, *2*, 85.
24.   Sherman, S. R.; Trampe, D. B.; Bush, D. M.; Schiller, M.; Eckert, C. A.; Dallas, A. J.; Li, J.; Carr, P. W. *Ind. Eng. Chem. Res.* **1996**, *35*, 1044.
25.   Carr, P. W. *Microchemical Journal* **1993**, *48*, 1.
26.   Kamlet, M. J.; Abboud, J.-L. M.; Abraham, M. H.; Taft, R. W. *J. Org. Chem.* **1983**, *48*, 2877.
27.   Abraham, M. H.; Whiting, G. S.; Doherty, R. M.; Shuely, W. J. *J. Chromatogr.* **1991**, *587*, 213.
28.   Li, J.; Zhang, Y.; Dallas, A. J.; Carr, P. W. *J. Chromatogr.* **1991**, *550*, 101.
29.   Abraham, M. H. *Chem. Soc. Rev.* **1993**, *22*, 73.
30.   Abraham, M. H.; Chadha, H. S.; Whiting, G. S.; Mitchell, R. C. *J. Pharm. Sci.* **1994**, *83*, 1085.
31.   Abraham, M. H.; Andonian-Haftvan, J.; Whiting, G. S.; Leo, A.; Taft, R. S. *J. Chem. Soc. Perkin Trans. 2* **1994**, 1777.
32.   Ikushima, Y.; Saito, N.; Arai, M.; Arai, K. *Bull. Chem. Soc. Jpn.* **1991**, *64*, 2224.
33.   Ely, J. F.; Haynes, W. M.; Bain, B. C. *J. Chem. Therm.* **1989**, *21*, 879.
34.   Van Alsten, J. G.; Eckert, C. A. *J. Chem. Eng. Data* **1993**, *38*, 605.
35.   Schmitt, W. J.; Reid, R. C. *Process Technol. Proc.* **1985**, *3*, 123.
36.   McHugh, M. A.; Seckner, A. J.; Yogan, T. J. *Ind. Eng. Chem. Res.* **1984**, *23*, 493.
37.   Reverchon, E.; Russo, P.; Stassi, A. *J. Chem. Eng. Data* **1993**, *38*, 458.
38.   Yau, J.-S.; Tsai, F.-N. *J. Chem. Eng. Data* **1993**, *38*, 171.
39.   Smith, V. S.; Ph.D. Thesis; Georgia Institute of Technology; **1995**.
40.   McEachern, D. M.; Sandoval, O.; Iñiguez, J. C. *J. Chem. Thermodyn.* **1975**, *7*, 299.
41.   Hansen, P. C.; Ph. D. Thesis; University of Illinois; **1985**.
42.   Bradley, R. S.; Cleasby, T. G. *J. Chem. Soc.* **1953**, 1690.
43.   Hansen, P. C.; Eckert, C. A. *J. Chem. Eng. Data* **1986**, *31*, 1.
44.   Macknick, A. B.; Prausnitz, J. M. *J. Chem. Eng. Data* **1979**, *24*, 175.
45.   Taylor, J. W.; Crookes, R. J. *J. Chem. Soc., Faraday Trans. 1* **1976**, *72*, 723.
46.   Johnston, K. P.; Ziger, D. H.; Eckert, C. A. *Ind. Eng. Chem. Fundam.* **1982**, *21*, 191.
47.   Coutsikos, P.; Magoulas, K.; Tassios, D. *J. Chem. Eng. Data* **1995**, *40*, 953.
48.   Kwiatkowski, J.; Lisicki, Z.; Majewski, W. *Ber. Bunsen Ges. Phys. Chem.* **1984**, *88*, 865.
49.   Colomina, M.; Jimenez, P.; Turrion, C. *J. Chem. Therm.* **1982**, *14*, 779.
50.   Schmitt, W. J.; Reid, R. C. *J. Chem. Eng. Data* **1986**, *31*, 204.
51.   Aihara, A. *Bull. Chem. Soc. Japan* **1959**, *32*, 1242.
52.   Burkhard, L. P.; Armstrong, D. E.; Andren, A. W. *J. Chem. Eng. Data* **1984**, *29*, 248.
53.   Radchenko, L. G.; Kitaigorodskii *Russ. J. Phys. Chem.* **1974**, *48*, 1595.
54.   McHugh, M. A.; Paulaitis, M. E. *J. Chem. Eng. Data* **1980**, *25*, 326.
55.   Suoqi, Z.; Renan, W.; Guanghua, Y. *J. Supercrit. Fluids* **1995**, *8*, 15.
56.   Bothe, H.; Cammenga, H. K. *1979* **1979**, *16*, 267.
57.   Johannsen, M.; Brunner, G. *Fluid Phase Equil.* **1994**, *95*, 215.
58.   Li, S.; Varadarajan, G. S.; Hartland, S. *Fluid Phase Equil.* **1991**, *68*, 263.

59. Jiménez, P.; Roux, M. V.; Turrrión, C. *J. Chem. Therm.* **1990**, *22*, 721.
60. Pouillot, F. L. L.; Ph.D. Thesis; Georgia Institute of Technology; **1995**.
61. Bradley, R. S.; Drury, T. *Trans. Soc.* **1959**, *55*, 1844.
62. de Kruif, C. G. *J. Chem. Therm.* **1980**, *12*, 243.
63. Burk, R.; Kruus, P. *Can. J. Chem. Eng.* **1992**, *70*, 403.
64. Sato, N.; Inomata, H.; Arai, K.; Saito, S. *J. Chem. Eng. Japan* **1986**, *19*, 145.
65. Hess, B. S.; Ph.D. Thesis; University of Illinois; **1988**.
66. Edwards, D. R.; Prausnitz, J. M. *J. Chem. Eng. Data* **1981**, *26*, 121.
67. Mitra, S.; Chen, J. W.; Viswanath, D. S. *J. Chem. Eng. Data* **1988**, *33*, 35.
68. Osborn, A. G.; Douslin, D. R. *J. Chem. Eng. Data* **1975**, *20*, 229.
69. Iwai, Y.; Mori, Y.; Hosotani, N.; Higashi, H.; Furuya, T.; Arai, Y.; Yamamoto, K.; Mito, Y. *J. Chem. Eng. Data* **1993**, *38*, 509.
70. Davies, M.; Malpass, V. E. *J. Chem. Soc.* **1961**, 1048.
71. Yau, J.-S.; Tsai, F.-N. *J. Chem. Eng. Data* **1994**, *39*, 827.
72. Sonnefeld, W. J.; Zoller, W. H.; May, W. E. *Anal. Chem.* **1983**, *55*, 275.
73. Sasse, K.; Jose, J.; Merlin, J.-C. *Fluid Phase Equil.* **1988**, *42*, 287.
74. Bartle, K. D.; Clifford, A. A.; Jafar, S. A. *J. Chem. Eng. Data* **1990**, *35*, 355.
75. Jones, A. H. *J. Chem. Eng. Data* **1960**, *5*, 196.
76. Ambrose, D.; Lawrenson, I. J.; Sprake, C. H. S. *J. Chem. Therm.* **1976**, *8*, 503.
77. Dobbs, J. M.; Wong, J. M.; Johnston, K. P. *J. Chem. Eng. Data* **1986**, *31*, 303.
78. de Kruif, C. G.; Smit, E. J.; Govers, H. A. J. *J. Chem. Phys.* **1981**, *74*, 5838.
79. Krukonis, V. J.; Kurnik, T. *J. Chem. Eng. Data* **1985**, *30*, 247.
80. Colomina, M.; Jiménez, P.; Roux, M. V.; Turrión, C. *J. Cal. Anal. Therm.* **1980**, *11*, 3.
81. de Kruif, C. G.; Van Ginkel, H. D. *J. Chem. Therm.* **1977**, *9*, 725.
82. Reverchon, E.; Donsì, G.; Gorgoglione, D. *J. Supercrit. Fluids* **1993**, *6*, 241.
83. Sako, S.; Shibata, K.; Ohgaki, K.; Katayama, T. *J. Supercrit. Fluids* **1989**, *2*, 3.
84. Colomina, M.; Jiménez, P.; Roux, M. V.; Turrión, C. *An. Quim.* **1986**, *82*, 126.
85. Tsai, K.-L.; Tsai, F.-N. *J. Chem. Eng. Data* **1995**, *40*, 264.
86. Iwai, Y.; Fukuda, T.; Koga, Y.; Arai, Y. *J. Chem. Eng. Data* **1991**, *36*, 430.
87. Bamberger, T.; Erickson, J. C.; Cooney, C. L.; Kumar, S. K. *J. Chem. Eng. Data* **1988**, *33*, 327.
88. Ambrose, D.; Lawrenson, I. J.; Sprake, C. H. S. *J. Chem. Therm.* **1975**, *7*, 1173.
89. Lamb, D. M.; Barbara, T. M.; Jonas, J. *J. Phys. Chem.* **1986**, *90*, 4210.
90. Aihara, A. *Bull. Chem. Soc. Japan* **1960**, *33*, 194.
91. Tan, C. S.; Weng, J. Y. *Fluid Phase Equilib* **1987**, *34*, 37.
92. Smith, N. K.; Stewart, R. C., Jr.; Osborn, A. G.; Scott, D. W. *J. Chem. Therm.* **1980**, *12*, 919.
93. Andon, R. J. L.; Biddiscombe, D. P.; Cox, J. D.; Handley, R.; Harrop, D.; Herington, E. F. G.; Martin, J. F. **1960**, 5246.
94. Iwai, Y.; Yamamoto, H.; Tanaka, Y.; Arai, Y. *J. Chem. Eng. Data*, **1990**, 35, 174.

Chapter 3

Phase Equilibria of Vegetable Oils with Near-Critical Fluids

Juan C. de la Fuente B., Tiziana Fornari, Esteban A. Brignole,
and Susana B. Bottini

PLAPIQUI, Universidad National del Sur—Consejo National de Investigaciónes
Cientificas y Tecnicas, C.C. 717, 8000 Bahia Blanca, Argentina

The application of the SRK equation of state for the prediction of the
phase behaviour of mixtures of vegetable oils with near critical solvents
is studied. The use of binary interaction parameters in the combinatorial
rules for both the co-volume and the energy parameter is discussed. Two
different sets of binary interaction parameters are needed in order to
correlate the vapor-liquid and liquid-liquid equilibria. This indicates a
serious limitation of van der Waals type of equations of state for
modeling the phase equilibria of this type of systems.

Vegetable oils are mixtures of triglycerides of saturated and unsaturated fatty acids;
i.e. mixtures of long-chain slightly polar molecules, with molecular weights on the
order of 850. Paraffinic solvents are used in the extraction and purification of
vegetable oils, hexane being the most common. Propane has also been used for the
purification of vegetable and fish oils. In recent times there has been a great interest in
the development of new processes for the extraction and purification of vegetable oils,
mono and diglycerides and fatty acids, using near critical fluids like carbon dioxide,
propane, ethane and their mixtures.

The design of these processes requires equilibrium data as well as methods for
the prediction and calculation of phase equilibria in mixtures of low molecular weight
near-critical fluids with vegetable oils. The thermodynamic modeling of oil-solvent
mixtures can be done on the basis of activity coefficient models or equations of state.
In a recent work, Fornari et al. (1) found that the combinatorial contribution to the
activity coefficients is the main source of nonideality. This is due to the large
differences in molecular size between solvent and oil molecules. These mixtures
exhibit partial liquid miscibility at high pressures and high solvent concentrations. In
the present work, taking into consideration the range of pressures and near-critical
conditions of interest, the application of the Soave-Redlich-Kwong, SRK (2) equation
of state is studied. The simultaneous modeling of liquid-liquid and vapor-liquid
equilibria is also considered.

Phase Behaviour of Vegetable Oils with Near Critical Fluids

Phase equilibrium data for propane with triglycerides have been presented by Coorens et al. (3) and de la Fuente et al. (4). Figure 1 shows a typical phase diagram for sunflower oil with propane. These systems exhibit negative deviations to Raoult's law (due to combinatorial or entropic effects) at low solvent concentration and slightly positive deviations in the solvent rich area. Above a certain temperature, i.e. the lower critical end point (LCEP), a region of partial liquid miscibility is reached. In the case of sunflower oil mixtures with ethane (Figure 2) the immisciblity region is wider and extends over all the studied pressure range (4). The possibility of having complete or partial miscibility in the liquid phase by moderate pressure changes, has some practical interest. The use of mixtures of ethane and propane as solvent, increases the miscibility gap with respect to that of pure propane, but retains the complete miscibility behaviour at higher pressures (Figure 3) (5). The partial miscibility phenomena is typical of mixtures with molecular size asymmetry, when the light component is near to its critical conditions (3-6). This behaviour can be identified as Type IV (Figure 4), following the classification of Van Konynenburg and Scott (7).

Thermodynamic Modeling

In order to apply the SRK equation of state it is necessary to estimate the values of the critical temperature (Tc), critical pressure (Pc) and acentric factor (ω) of the sunflower oil. The critical temperature of sunflower oil was estimated using Fedors' group contribution method (8). The molar composition of sunflower oil is the following (acid base): palmitic 6.7%, stearic 3.34%, oleic 25.83%, linoleic 63.91% and linolenic 0.22%. The Pc and ω values were adjusted, with the estimated Tc held constant, so that the SRK EOS optimally correlated the liquid density (9) and vapor pressures (10) of the pure oil. The values of the estimated parameters for sunflower oil are given in Table I. Also included in this table are the corresponding parameters for ethane and propane.

Table I
Critical parameters and acentric factor for ethane, propane and sunflower oil

	Tc (K)	Pc (bar)	ω
Ethane	305.4	48.8	0.091
Propane	369.8	42.5	0.145
Sunflower oil	1042.4	8.2	0.714

The following mixing and combination rules were used to calculate the co-volume and energy parameter of the mixtures:

$$b = \sum \sum x_i\, x_j\, b_{ij} \qquad \text{with} \qquad b_{ij} = (1 - \text{kij}(b))\, (b_i + b_j)\,/\,2 \qquad (1)$$

$$a = \sum \sum x_i\, x_j\, a_{ij} \qquad \text{with} \qquad a_{ij} = (1 - \text{kij}(a))\, (a_i\, a_j)^{\frac{1}{2}} \qquad (2)$$

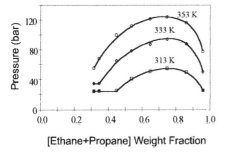

Figure 1. Phase - equilibrium isotherms for fluid mixtures of Sunflower Oil - Propane (*4*). (●): 313 K; (■): 353 K.

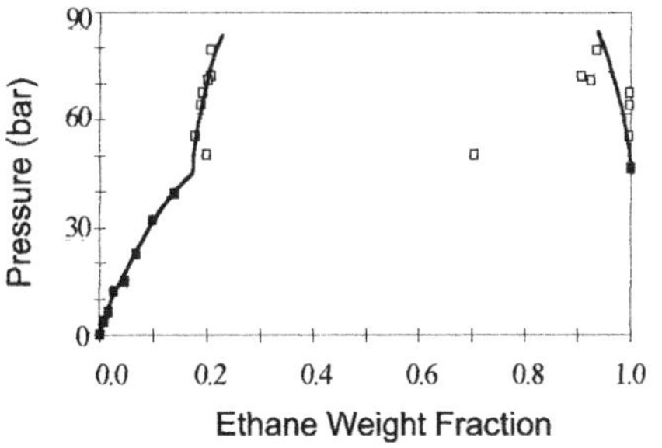

Figure 2. Vapor-liquid (VLE) and liquid - liquid (LLE) equilibrium data for the Ethane-Sunflower Oil system at 303.2 K (*4*). (■) VLE; (□) LLE.

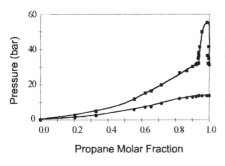

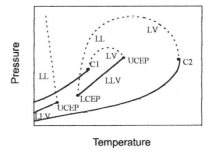

Figure 3. Vapor-liquid (VLE) and liquid - liquid (LLE) equilibrium data for the Ethane - Propane - Sunflower Oil system (*4*). Propane - Ethane molar ratio: 2.07. (■) VLE; (□) LLE.

Figure 4. Pressure - temperature projection of type IV systems (*7*).

In most applications of cubic equations of state, it is common to assume that the interaction parameter for the co-volume (kij(b)) is zero. However, this assumption leads to a poor correlation of the vapor pressure of binary solutions of propane or ethane with sunflower oil, as can be seen in Figure 5. The introduction of a binary parameter in b greatly improves the correlation of vapor pressures of light hydrocarbon - sunflower oil mixtures. However, if this kij(b) value is used in the region of partial liquid miscibility, the equation of state is not able to correlate (Figure 6) or predict, even for the solvent supercritical conditions (Figure 7), the high pressure liquid-liquid equilibria. The extended Maxwell Equal Area Rule (*11*) algorithm was used to calculate the liquid-liquid equilibrium conditions. The a and b parameters were adjusted so as to minimize the errors in liquid compositions. The kij(b) and kij(a) values for vapor-liquid (VLE) and liquid-liquid (LLE) equilibria of sunflower oil binary mixtures with ethane or propane, are given in Table II. It can be seen that, whereas the values of the energy interaction parameters kij(a) are similar, the numerical values of the co-volume parameter required to fit vapor - liquid or liquid - liquid equilibria are completely different. When the kij(b) parameter obtained from fitting liquid-liquid equilibrium data is used for the prediction of vapor pressures, the agreement with experimental data is poor for mixtures with propane weight fractions in the liquid phase below 0.2 (Figure 6). These results indicate a serious limitation of van der Waals type of equations of state for the prediction of vapor-liquid and liquid-liquid equilibria with a single set of parameters.

Table II
Binary interaction parameters for the SRK equation of state

System	Type of data	b linear kij(a)	b quadratic kij(a) / kij(b)
Sunflower oil + ethane	VLE	-0.063	0.014 / 0.033
	LLE		0.080 / 0.220
Sunflower oil + propane	VLE	-0.050	0.034 / 0.044
	LLE		0.040 / 0.210

The LLE binary interaction parameters were used to estimate the pressure that would be required to obtain complete miscibility in the system ethane-sunflower oil. As can be seen in Figure 8, that pressure would be above 1000 bar.

Reasonable agreement with the available experimental information on ternary mixtures of ethane - propane - sunflower oil (5) is obtained when the binary liquid-liquid interaction parameters from Table I are used (see Figure 9).

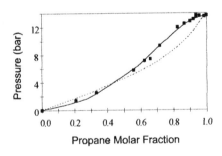

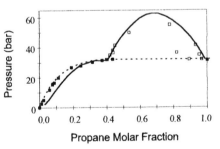

Figure 5. Vapor pressures of Propane in mixtures with Sunflower Oil at 313 K. (■): Experimental ; (----) SRK, kij(a)=-0.05 and kij(b)=0; (—) SRK, kij(a)=0.034 and kij(b)=0.044.

Figure 6. SRK predictions of phase equilibria for Sunflower Oil - Propane mixtures at 353 K. Experimental: (■)VLE, (□)LLE; SRK: (----) kij(a)=0.034 and kij(b)=0.044; (—) kij(a)=0.040 and kij(b)=0.210.

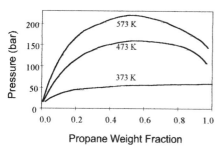

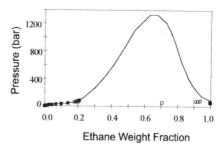

Propane Weight Fraction

Ethane Weight Fraction

Figure 7. SRK predictions of phase equilibria for the Propane - Sunflower Oil system at high pressures, using vapor - liquid interaction parameters.

Figure 8. SRK predictions of high pressure liquid - liquid equilibria of Ethane - Sunflower Oil mixtures at 303.2 K.

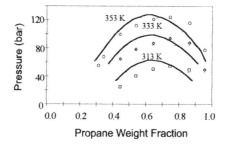

Propane Weight Fraction

Figure 9. Experimental and predicted liquid - liquid equilibria for Ethane - Propane - Sunflower Oil mixtures. Propane - Ethane molar ratio: 2.07. (□): Experimental; (—): SRK.

Discussion and Conclusions

Propane and ethane and their mixtures are very good solvents for the extraction of vegetable oils under near critical conditions. Propane - sunflower oil mixtures are completely miscible at ambient temperature. At temperatures closer to the propane critical point there is a region of partial miscibility. By a pressure increase at a given temperature, the complete miscibility condition can be recovered. Ethane at near critical conditions depicts only partial miscibility within the pressure range investigated. The use of mixtures of ethane and propane modifies the miscibility gap at a given temperature, and complete miscibility can be achieved by moderate variations in pressure and/or temperature.

The high pressure vapor-liquid equilibrium data are well correlated with cubic equations of state using binary interaction parameters for both the co-volume and the energy parameter. However, the correlation of vapor-liquid and liquid-liquid equilibria is not possible with the same set of parameters. The correlation of liquid-liquid equilibria requires a much higher value of the binary coefficient for the co-volumen parameter. This indicates a serious limitation of the Van der Waals type of equations of state for modeling the phase equilibria of this type of systems.

Literature Cited

(1) Fornari T., Brignole E. A. and Bottini S., Application of UNIFAC to Vegetable Oil - Alkane Mixtures, *J. Am. Oil Chem. Soc.* **1994**, *71* (4), 391.

(2) Soave G., *Chem. Eng. Sci.* **1972**, *27*, 1197.

(3) Coorens H. G. A., Peters C. J. and de Swaan Arons J., Phase Equilibria in Binary Mixtures of Propane and Tripalmitin, *Fluid Phase Equilibria*, **1988**, *40*, 135.

(4) de la Fuente B. J.C., Ph.D. dissertion, Universidad Nacional del Sur, Bahía Blanca, **1994**.

(5) de la Fuente B. J.C., Mabe G., Bottini S. and Brignole E. A., Phase Equilibria in Binary Mixtures of Ethane and Propane with Sunflower Oil, *Fluid Phase Equilibria*, **1994**, *101*, 247.

(6) Rowlinson J.S. and Swinton F.L., Liquids and Liquid Mixtures, 3th edition, Butterworth, London, **1982**.

(7) Van Konynenburg P.H. and Scott R. L., *Phil. Tran. Roy. Soc.* **1980**, *298*, 495.

(8) Reid R., Prausnitz J. M. and Poling B., The Properties of Gases and Liquids, 4th edition, Mc Graw-Hill Book Company, New York, **1987**.

(9) Bailey A. E., *Industrial Oil and Fat Products*, Intersc. Publishers Inc., New York, **1951**.

(10) Perry E. S., Weber W. H. and Daubert B. F., Vapor Pressures of Phlegmatic Liquids. I. Simple and Mixed Triglycerides, *J. Am. Chem. Soc.* **1949**, *71*, 3720.

(11) Eubank P.T. and Hall K. R., Equal–Area Rule and Algorithm for Determining Phase Compositions, *AICHE Journal* **1995**, *41* (4), 924.

Chapter 4

Supercritical Fluid Extraction of Recycled Fibers: Removal of Dioxins, Stickies, and Inactivation of Microbes

Carol A. Blaney and Shafi U. Hossain[1]

Kimberly-Clark Corporation, Long Range Research and Development, 1400 Holcomb Bridge Road, Roswell, GA 30076

Supercritical carbon dioxide and supercritical propane were found to be effective solvents in extracting stickies and trace chlorinated organics, including dioxins, from recycled fibers. These undesirable components are not effectively removed with current recycled fiber processing techniques. It was also found that endogenous yeast and mold spores on the fibers were inactivated with supercritical carbon dioxide. An economic analysis was performed for a recycled paper pretreatment process which utilizes semi-batch supercritical fluid extraction. Matrix effects were ignored owing to lack of data. Results are encouraging, estimating cost ranges of 7-17 cents per pound of fibers treated -- costs well within reasonable price targets to pretreat premium paper and tissue products.

Efficient management of solid waste streams, of which cellulose-based materials such as waste paper constitute a significant part (roughly 40%), represents a major technological challenge. In recent years, an impressive array of new technologies has emerged which addresses problems of recycling waste paper. Novel screening systems and sophisticated flotation techniques have emerged which have, in large measure, successfully addressed the problem of deinking printed stock. Bleaching sequences which avoid the use of chlorine or chlorine compounds, and rely upon hydrogen peroxide, dithionites, or formamidine sulfinic acid for attaining acceptable levels of brightness, are also under investigation in various research laboratories around the world. However, the technologies mentioned do not effectively address the problem of stickies, dioxins, and microbes in the waste paper feedstock.

Recycled Fiber Contaminants

Stickies. One of the problems in the use of recycled fibers is due to sticky contaminants consisting mostly of organic adhesives and tackifiers used in the converting process, such as styrene-butadiene rubbers, acrylates, and polyvinyl

---

[1]Current address: N7753 Sundown Court, Sherwood, WI 54169

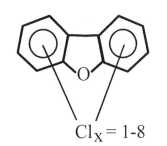

polychlorinated
dibenzo-para-dioxins
(PCDDs or "dioxins")

2,3,7,8-tetrachloro-
dibenzo-para-dioxin
(reportedly the most
toxic congener)

polychlorinated
dibenzofurans
(PCDFs, or
"dibenzofurans")

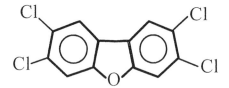

2,3,7,8-tetrachloro-
dibenzofuran
(reportedly the most
toxic congener)

Fig. 1. Structure of Dioxin and Related Compounds.

acetate. When waste papers containing these adhesives/tackifiers are defibered, stickies are broken down into various sizes ranging from 0.05 to 0.5 mm. These stickies cause off-quality paper and paper machine downtime, sometimes requiring the use of environmentally hazardous solvents to clean equipment fouled by deposition of stickies on wire, felts, presses, rolls, and drying cylinders. Current methods to deal with stickies (additives, separation processes, and processing at high-temperatures, for example) are not sufficient to completely eliminate the problems associated with stickies in the feedstock. Eliminating stickies via furnish selection has proven unreliable and impossible to implement.

Part of the experimental work presented in this paper addresses stickies problems via extracting recycled fibers with supercritical fluid, before subjecting the fibers to subsequent deinking and bleaching operations.

Dioxins. One aspect of waste paper reuse is concerned with the probable presence of small quantities of reportedly toxic compounds such as "dioxins" (see Fig. 1) and similar chlorinated organic compounds in waste papers, which has received much attention recently in the news media, and has caused considerable public anxiety. Kraft pulps, when bleached with sequences including an elemental chlorine stage, sometimes contain small but detectable levels of dioxins and related chlorinated organic compounds which may be perceived by the consuming public as a health hazard. Bleached Kraft fibers under a variety of guises (e.g. coated paper, ledger paper, etc.) are often present in substantial quantities in waste paper purchased from commercial dealers. Clapp and Truemper published AOX levels in paper made from recycled fibers (1), indicating many recycled fiber feedstocks contain up to 1200 mg/kg (ppm) AOX, where AOX refers to the broad group of halogenated organics, of which dioxin and dibenzofurans are a subset.

Current state-of-the-art recycled paper processing methods, which include novel screening systems and sophisticated flotation techniques, do not address removal of dioxins or related chlorinated organic compounds. The present work attacks the problem by demonstrating the removal of dioxin from the waste paper feed stock via supercritical fluid extraction *before* it is subjected to conventional deinking/bleaching operations.

Microbes. The question of cleanliness and sanitation also arises when using post-consumer recycled paper products. It is therefore desirable to inactivate any potentially pathogenic organisms in the feedstock in the event that bleaching steps such as ozonation (which are presumed to sanitize) are not used. This work indicates that supercritical fluid treatment may provide a pathway towards inactivation of microbial pathogens.

Experimental Section

The experimental work done was designed to ascertain, in a definitive manner, if carbon dioxide or propane, under various conditions, are suitable solvents for the extraction of dioxins and stickies from secondary fibers, as well as whether these solvents are able to inactivate microbes. The quest for an unambiguous and

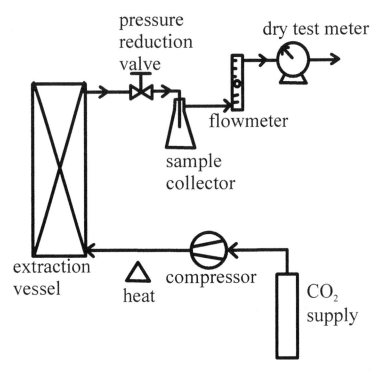

Fig. 2. Supercritical Fluid Extraction Apparatus.

Table 1. Stickies Content of Pulp Fibers before and after SC-CO2
Extraction (Determined by Soxhlet Extraction) and Calculated
Extraction Efficiency (% Removal).

Soxhlet Extr. Solvent:	Control (before SCFE)	"Product" (after SCFE)	%Stickies Removed via SCFE
EtOH/Bz	1.3	0.47	64%
Acetone	0.86	0.32	63%
CH2Cl2	1.2	0.25	79%
Hexane	1.0	0.45	55%

expeditious answer to the question stressed the need for economy of effort and simplicity of execution in the design of the experimental program.

Dioxin Extraction Studies. Recycled paper was selected to represent both softwood and hardwood fibers and included white and colored ledger grades and coated sulfate (magazine) papers. The papers were ground using a Wiley Mill to particle sizes of less than 0.5 mm, and subsequently spiked with 232 ppt (parts per trillion, e.g., grams per trillion grams) "dioxin".

Supercritical carbon dioxide solvent extraction conditions were as follows: 71 degrees C and 34.5 Mpa (5000 psi). Extraction efficiencies were measured using a 3-liter batch extractor depicted in Fig. 2, and a supercritical carbon dioxide solvent-to-feed ratio of 105 (grams of solvent per gram of recycled fiber). State-of-the-art high resolution GC-MS was used to analyze the dioxin in the samples, and 5 replicates were performed for both the 'feed' and the extracted 'product'. Extraction efficiencies were greater than 95% (dioxin removal).

Similar experiments were performed using supercritical propane at 125 degrees C and 34.5 MPa (5000 psi), with a solvent-to-feed ratio of 30. In this experiment, however, the feed sample was spiked with 232 ppt $^{13}C_{12}$ (labeled) dioxin, and 'feed' and extracted 'product' samples were analyzed for both native (unlabeled) and spiked (labeled) dioxin.

It was found that the supercritical propane removed 95% of the spiked dioxin, and only 85% of the native dioxin. This is not surprising. It suggests that the native dioxin is more sterically trapped inside the fiber interstices, and/or that the binding energy of the older, native dioxin is stronger than for the spiked dioxin. This finding is in agreement with Steve Hawthorne of the University of North Dakota (2). Hawthorne found that spiked naphthalene in sludge was extracted faster than naphthalene in aged sludge contaminated with polyaromatic hydrocarbons.

Stickies Experiments. Representative paper fibers containing stickies were used in these extraction experiments, and included ice cream cartons, paper pads with adhesive bindings, and book bindings. The paper was ground in a Wiley Mill to particle sizes of less than 0.5 mm in size. In a procedure similar to the experiments described above, the paper was extracted with carbon dioxide at 60 degrees C, 34.5 MPa (5000 psi), and a solvent-to-feed ratio of 30. Standard Soxhlet extractions were performed to determine percent extractables in the 'feed' and the extracted 'product'. The Soxhlet extraction solvents used were hexane, methylene chloride, acetone, and ethanol/benzene. Extraction efficiencies for stickies removal ranged from 64% to 79%, depending on the Soxhlet extraction solvent used. Detailed results for individual Soxhlet extractions, as well as calculations of extraction efficiencies, are shown in Table 1. Table 2 compares the extraction efficiencies of supercritical carbon dioxide with those of supercritical propane.

Similar experiments to supercritical carbon dioxide extraction were performed with supercritical propane at 125 degrees C and 34.5 MPa (5000 psi), and a solvent-to-feed ratio also of 30. Stickies extraction efficiencies were 69% to 91%, depending on the Soxhlet extraction solvent used. Results for individual soxhlet extractions are summarized in the middle column of Table 2.

Another experiment using supercritical propane at 125 degrees C and only 8.26 MPa (1200 psi) gave stickies extraction efficiencies of 42 to 70 percent. Results are summarized in the right-hand column of Table 2.

IR spectra of Soxhlet extraction residues show hexane only removed the natural wood extractives such as resins and fatty acids. All other solvents (and solvent blends) used showed styrene-butadiene, polyvinyl acetate and other representative stickies in the extract.

Microbe Experiments. Similar experiments were performed extracting recycled fibers with carbon dioxide at 60 degrees C, 34.5 MPa (5000 psi), and a solvent-to-feed ratio of 30 (20 minute exposure time). Analysis of 'feed' samples and extracted 'product' samples confirmed inactivation of all endogenous yeast and mold spores, which is not surprising since CO_2 is a waste product of these microbes.

Economics

A detailed engineering cost analysis (CHEMCAD computer-aided design system) was performed on a solvent batch extraction process, comparing costs of using supercritical carbon dioxide as the solvent, supercritical carbon dioxide with an entrainer as the solvent, and supercritical propane as the solvent.

Assumptions. The plant design was for a 100 Bone Dry Ton Per Day (BDTPD) recycled pulp/paper facility. It was assumed that the extractions were solubility-limited (matrix effects were ignored due to a limited amount of experimental data from the extraction experiments). A 1% solvent loss rate per circulation was assumed, which is probably higher than an actual plant would run, and hence will predict costs on the high side. Note that operating temperatures and pressures were not yet optimized for this analysis, which also would tend to predict costs on the high end. It was assumed that the use of an ethanol co-solvent at concentrations of 5% ethanol in carbon dioxide would give an entrainer effect of 3. Whether this assumption is valid is debatable; however it is not unusual to experience entrainer effects much higher than 3 in such systems, especially if the solutes have polar groups which would interact with the polar hydroxyl group on ethanol. The cost of drying the fibers prior to the use of the carbon dioxide-plus-ethanol solvent (to prevent azeotrope formation of ethanol-water) was assumed to be relatively small.

Propane was assumed to have an enhancement factor roughly 2 to 3 times higher than for CO_2 at the same reduced temperature and density. Since the vapor pressure of the solute is likely to be about 20 times as high at the temperatures encountered for propane, the solubility in propane was given a multiplicative factor of 50 with respect to the CO_2 estimated solubilities.

A semi-batch process was used in which the supercritical fluid was continuous, and the solids were loaded into several high tensile strength steel extraction vessels in parallel, processed, and then unloaded. The SCF stream was reduced in pressure in a separate blowdown vessel set at 5.2 MPa (750 psi), and the SCF recycled. A purge compresser was provided to remove residual SCF from the extraction vessels prior to opening, returning it to the main compressor at 5.2 MPa. An option was included for a granular activated carbon polishing step at the lower pressure to reduce contaminant levels in the SCF recycle stream.

Results of Economic Analysis. A supercritical carbon dioxide semi-batch extraction facility, without the use of an entrainer, shows an initial capital investment of $74MM with an ongoing operating cost of $0.33/pound of bone dry fibers. Addition of a 5% ethanol entrainer to increase the solubilities by a factor of 3 requires a lower solvent-to-feed ratio, bringing costs down significantly and giving an initial capital investment of $40MM, with an operating cost of $0.17/pound of bone dry fibers.

Using supercritical propane as a solvent allows the batch process to be run at much lower pressures. Since propane is flammable, provision was made for flushing the extraction vessels with nitrogen during loading and unloading to prevent the formation of any explosive mixture. The propane is then flared. Only when the mixture is beyond any possibility of explosion would the vessel be opened. Similarly on closing the vessel, the same process would be done in reverse to remove oxygen. This process had the lowest cost of the three discussed here, with an initial capital investment of $13MM, and $0.067/lb operating costs.

The results of the economic study are given in Tables 3 and 4, and Figure 3 is a further condensed summary of the results in a bar chart.

Summary and Conclusions

SCFE was effective in removing the majority of dioxins (85-95%) as well as the majority (64-91%) of stickies. For stickies removal, SC-CO$_2$ at 34.5 MPa performed better than SC-propane at 8.25 MPa and worse than SC-propane at 34.5 MPa. Initial data show inactivation of common yeast and mold spores during supercritical CO$_2$ extraction. Preliminary economics indicate that operating costs for a system using carbon dioxide with 5% ethanol co-solvent may cost nickels per pound of fibers treated, whereas liquid propane, pennies per pound. The associated fire hazards with the use of propane make carbon dioxide-based systems potentially more attractive.

The authors would like to stress that, while the results are promising, the work presented here represents the initial stages of a research program. With refinement of techniques and imaginative modification of the basic supercritical solvents, both the extraction efficiencies and the process economics are likely to improve.

Recommendations

The authors recommend the following research program as the next step in developing a viable, economical commercial process for pretreatment of recycled fiber feedstock with supercritical fluid extraction:

Table 2. Comparisons of Extraction Efficiencies of Supercritical
Solvents (Determined by Soxhlet Extraction of Sample Before and
After SCFE).

Soxhlet Extr. Solvent:	SC-CO2 34.5MPa 60 deg.C	SC-Prop 34.5MPa 125 C	SC-Prop 8.26MPa 125 C
EtOH/Bz	64%	69%	42%
Acetone	63%	86%	74%
CH2Cl2	79%	91%	70%
Hexane	55%	98%	93%

Table 3. Calculated Costs of a 100 BDTPD Semi-Batch Extraction
Facility to Pretreat Recycled Fibers via Supercritical Carbon Dioxide,
with and w/o an Ethanol Co-Solvent.

Operating Pressure	10.4MPa (1500psi)	13.8MPa (2000psi)	20.7MPa (3000psi)
Pure CO2, capital costs	$89MM	$74MM	$83MM
Pure CO2, cents per pound	37	33	37
CO2+5% EtOH, capital costs	$43MM	$40MM	$44MM
CO2+5% EtOH, cents per pound	18	17	19

Table 4. Calculated Costs of a 100 BDTPD Semi-Batch Extraction
Facility to Pretreat Recycled Fibers via Supercritical Propane.

Operating Pressure	10.4MPa 1500psi	13.8MPa 2000psi	20.7MPa 3000psi
Propane, 102 C, capital costs	$13MM	$15MM	$16MM
Propane, 102 C, cents per pound	6.7	7.0	7.4
Propane, 121 C, capital costs	$13MM	$14MM	$15MM
Propane, 121 C, cents per pound	6.4	7.0	7.4

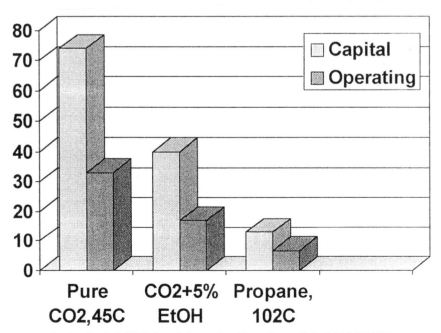

Fig. 3. Capital ($MM) and Operating Costs (cents/lb) of 100 BDTPD
Supercritical Fluid Extraction Plant.

- Evaluate the effect of entrainer(s) on enhancing removal of stickies and chlorinated organics such as dioxins.
- Investigate the ability of supercritical conditions to inactivate spiked pathogens such as hepatitis and strep.
- Obtain design data for solubilities, binding energies, and partitioning kinetics of stickies (styrene-butadiene, acrylates, and polyvinyl acetate) and dioxins (or model compounds which are less toxic but mimic dioxin in chemical nature).
- Reassess economics with revised data.

Kimberly-Clark Corporation holds four U.S. Patents covering this work (3-6).

Acknowledgments

The authors are grateful for the assistance provided by Phasex Corporation in Lawrence, Massachusetts in conducting the extraction experiments, and for Dr. Charles A. Eckert in the Dept. of Chemical Engineering at the Georgia Institute of Technology, Atlanta, Georgia, for his assistance in conducting the economic analysis.

Literature Cited

1) Hawthorne, S., University of North Dakota, as reviewed in LC-GC *vol. 13*, no. 7, July **1995**, pp 542-553 by Ronald E. Majors.

2) Clapp, R. T., Truemper, C. A., Aziz, S., and Reschke, T., *Tappi Journal*, vol. *79*, no. 3, pp. 111-113, March **1996**.

3) Blaney, C. A., and Hossain, S. U., US Patent No. 5,074,958, "Method For Removing Polychlorinated Dibenzodioxins And Polychlorinated Dibenzofurans And Stickies From Secondary Fibers Using Supercritical Propane Solvent Extraction".

4) Blaney, C. A., and Hossain, S. U., US Patent No. 5,009,746, "Method For Removing Stickies From Secondary Fibers Using Supercritical CO_2 Solvent Extraction".

5) Hossain, S. U., and Blaney, C. A., US Patent No. 5,009,745, "Method For Removing Polychlorinated Dibenzodioxins And Polychlorinated Dibenzofurans From Secondary Fibers Using Supercritical CO2 Extraction".

6) Hossain, S. U., and Blaney, C. A., US Patent No. 5,213,660, "Secondary Fiber Cellulose Product With Reduced Levels Of Polychlorinated Dibenzodioxins And Polychlorinated Dibenzofurans".

NATURAL MATERIALS

Chapter 5

Meeting the Natural Products Challenge with Supercritical Fluids

D. E. Raynie

Miami Valley Laboratories, Procter & Gamble Company, P.O. Box 538707, Cincinnati, OH 45253–8707

The unique nature of supercritical fluids is discussed relative to the needs inherent in extracting materials from natural products. An overview of the properties of supercritical fluids and the implications of these properties to achieve high mass transfer rates and enhance extraction processes is given. A description of the supercritical fluid extraction process and applications to natural products is presented.

The removal of the components of natural products from their source, whether for analytical or processing purposes, is historically one of the oldest chemical separation problems. Predating even the alchemists, the production of dyes, perfumes, and foods from natural products, as well as alcohol fermentation processes, exemplify the economic importance of natural products separations on the civilization of mankind. Despite this historical importance, the development and improvement of these processes has been relatively slow and incremental. Perhaps notable among these process improvements are fractional distillation methods, counter-current distribution processes, and chromatographic procedures.

Separations employing supercritical fluids represent a discontinuous breakthrough for natural products applications -- a generation of separations which do not rely on traditional (ambient) liquids. When temperatures and pressures approach the critical point, physical properties become more conducive to mass transfer. (Note that some definitions require temperatures and pressures above the critical point. In practice, conditions **near** the critical point can provide the physical properties exploited for supercritical fluid extraction (SFE).) For example, supercritical fluids, compared with liquids, possess diffusion, viscosity, and surface tension properties that are more gas-like, while having liquid-like density and solvating power. Furthermore, these properties can be altered through subtle changes in temperature and pressure.

68

Table I displays a comparison of some physical properties of gases, liquids, and supercritical fluids. As shown in the Table, supercritical fluids have densities on the same order as liquids, yet their diffusion rates tend to be an order of magnitude faster than liquids. Meanwhile, like gases, supercritical fluids do not have any surface tension and have similar viscosities. These properties combine to form a unique medium in which to perform natural product extraction processes. In fact, several years ago, some researchers used the term "destraction" when discussing SFE to highlight the similarity of the technique to both (volatility-based) distillation and (solubility-based) extraction.

Table I.
General Ranges of Selected Physical Properties of Gases,
Supercritical Fluids, and Liquids

	Density $(g\ mL^{-1})$	Diffusivity $(cm^2\ s^{-1})$	Viscosity $(g\ cm^{-1}\ s^{-1})$	Surface Tension $(dynes\ cm^{-1})$
Gas	$(0.6\text{-}2) \times 10^{-3}$	0.1-0.4	$(1\text{-}3) \times 10^{-4}$	0
Supercritical Fluid	0.2-1.0	$(2\text{-}7) \times 10^{-4}$	$(1\text{-}9) \times 10^{-4}$	0
Liquid	0.6-1.6	$(0.2\text{-}2) \times 10^{-5}$	$(0.2\text{-}3) \times 10^{-2}$	30-60

Extraction Basics

The importance of these solvent physical properties related to extraction can be emphasized through an understanding of the general extraction process. Figure 1 demonstrates the processes involved in the extraction of a compound (depicted as "X") from a solid sample matrix. For extraction to occur, the extracting solvent (regardless of its state) must diffuse into the sample, solubilize the compound of interest, and diffuse back through the sample particle. This process is dependent on the rate of diffusion through the solid sample, the wettability of the sample pores, the solubility of the compound of interest in the solvent, and the partitioning and adsorption of the extracted compound between the sample and the solvent, as well as sample specific properties like particle size and porosity. Once transported to the sample surface, the compound of interest must overcome the energy of adsorption at the particle surface and be swept away from the sample in the bulk solvent. Thus, the solvating ability (related to solvent density) and the ability to penetrate the sample (related to diffusion, viscosity, and surface tension) play key roles in the extraction process. Of course, differences in the extraction procedures will become evident based on the specific objective, *e. g.*, analytical or process-scale, but in all general cases, the favorable properties of supercritical fluids tends to enhance extractions compared with liquid-based procedures.

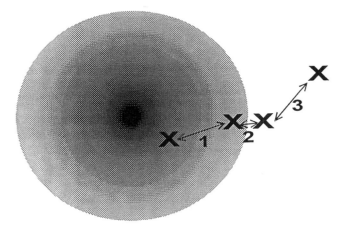

Figure 1. Schematic of the generalized extraction process. Steps include: 1) solubilization of the compound to be extracted and transport through the sample particle, 2) transport of the extracted compounds from the particle surface to the bulk extracting solvent, and 3) transport of the extracting solvent (with the extracted material) away from the sample.

Extraction processes, as described above, can be modeled, for example, with the "hot ball" model (1). This model assumes a solid matrix containing small quantities of extractable compounds which diffuse out of the homogeneous spherical particle into the extraction medium, where the extracted compounds are infinitely dilute. With these considerations, the extraction rate is dependent on the mass transfer out of the matrix, rather than solubility. This rate is obtained through the expression for the ratio of the mass, m, of extractable material remaining after time t to the initial mass of extractable material, m_o,

$$m/m_o = (6/\pi^2)\Sigma(1/n^2)\exp(-n^2\pi^2Dt/r^2)$$

where n is an integer, D is the diffusion coefficient of the material in the sample matrix, and r is the radius of the spherical sample. This equation reduces to a sum of exponential decays and a plot of $\ln(m/m_o)$ versus time becomes linear at extended time intervals. The physical explanation of the model is that during the initial phases of an extraction, there is a concentration gradient at the surface of the sphere and diffusion from the sphere is rapid. As the extraction continues, the concentration gradient near the surface continues to be large and the rate of diffusion is proportional to the concentration gradient. Ultimately the concentration across the entire sphere becomes even and the rate of diffusion (and, hence, extraction) is a simple exponential decay. While all extractions from solid matrices will undergo this type of diffusion-based extraction, in practice, limitations due to solute solubility in the extraction fluid are sometimes observed. This is especially true for analytical-scale extractions (to be discussed) where there is a relatively large mass of extractable material. Hawthorne *et al.* propose the use of kinetic plots (*i. e.,* graphs of extraction yield as a function of time) to distinguish diffusion- (or kinetic) limited extractions as opposed to solubility-limited extractions (2). Solubility-limitations can be overcome by adjusting the sample size or amount of extracting fluid (increasing either the dynamic flow rate or the extraction time), or increasing the solvating ability of the extraction fluid.

Use of Supercritical Fluids for Extraction

SFE processes are conceptually very simple and offer a wide range of flexibility in how they are performed. As a general rule, SFE operates as follows: The sample is placed into a closed vessel capable of withstanding the operational conditions of elevated temperature and pressure. (In many cases, the sample is a solid. Liquid samples can be directly accommodated, may flow in a counter-current fashion, or may be immobilized on a support material.) The extracting fluid is pressurized and delivered by a pump through the thermostated extraction vessel. The fluid may extract in either a dynamic flow-through process or may "soak" the sample in a static manner. After leaving the extraction vessel, the fluid (carrying the extracted sample) flows through some type of pressure reduction region into a collection device. In practice, pressure gauges, flow monitors, isolation valves, and other ancillary devices may be included in the system, and the extracting fluid may return to the sample vessel via a recycle mode.

Several advantages stem from the use of supercritical fluids for extraction processes:

- **Selectivity** -- Because the solvating ability of supercritical fluids can be altered through changes in temperature and/or pressure, SFE has the potential to preferentially dissolve and extract selected classes of compounds. This advantage may also lead to extracts which are "cleaner" or more pure.
- **Speed and Efficiency** -- As discussed, the mass transfer limits to extraction are reduced due to the rapid diffusion in supercritical fluids and the absence of surface tension that allows for better penetration and wetting of sample pores.
- **Oxygen-free Environment** -- SFE takes place in sealed vessels which can be devoid of oxygen, minimizing the potential of sample oxidation. Other types of sample degradation can also be avoided. For example, the typical steam distillation of flavor and fragrance compounds can lead to sample hydrolysis, which can be avoided with SFE using carbon dioxide.
- **Post-Extraction Manipulation** -- Depending on the fluid chosen, SFE can eliminate the need for post-extraction solvent evaporation, can e directly coupled to a variety of analytical techniques, or may be utilized in a more unique manner, such as the crystallization of pharmaceuticals.

- **Low Temperatures** -- Many of the commonly used supercritical fluids have critical temperatures less than 100 °C and some (*e. g.*, carbon dioxide, ethane, ethylene, trifluoromethane, sulfur hexafluoride) are less than 50 °C. This allows the extraction of thermally unstable, and perhaps volatile, materials.

While the advantages of SFE are significant, two major disadvantages have slowed the growth and acceptance of the field. SFE has a high cost, though the cost per extraction may be favorable. The implications of this cost are discussed later. Also, because SFE is still a developing field, the knowledge base available for developing and evaluating new applications is also developing. However, as SFE finds wider acceptance, this knowledge base will expand and this limitation will diminish. Safety concerns are important with SFE, though they should not be viewed as a barrier to its use. Implementation of standards, such as those issued by the ASTM, in the development of commercial systems and good laboratory and/or engineering practices alleviate most of these concerns.

The most commonly used supercritical fluid is carbon dioxide. The reasons for its use and the resulting advantages of this fluid are several. Carbon dioxide is inexpensive and widely available with a high level of purity. Its critical parameters (31 °C and 1070 psi) are convenient to use. The low critical temperature leads to the advantages stated above, but in practice, temperatures up to about 200 °C are not uncommon. The upper pressure range (typically around 5000 psi for process operations and 10,000 psi for analytical-scale extractions) is easily attained with conventional pumping systems. Importantly, carbon dioxide is a nonflammable inorganic fluid. Consequently the concerns over organic solvent use, such as residues, handling and exposure, disposal, and product contamination, are mitigated.

The preceding discussion has been general for all types of extraction. One means of classifying extraction types is based on the objectives of the extraction. Extractions being done as a preliminary sample preparation step for subsequent chemical or physical analysis (analytical-scale) and extractions performed for the isolation of material for subsequent processing (process-scale) can by approached differently based on the overall goal. These differences may include:

- **Scale** -- Analytical extractions require samples that are representative of the system under study and meet the sensitivity requirements of the specific analysis. Thus, several milligrams to tens of grams are used. On the other hand, process-scale extractions may use several kilograms.
- **Purity** -- While one of the goals of all extractions is the separation of the material of interest from the bulk sample, the desired purity may vary depending on the availability of subsequent clean-up procedures.
- **Operations** -- Analytical-scale extractions are generally performed batchwise, whereas continuous systems have been developed for process operations.
- **Solvent-to-Feed Ratio and Phase Behavior** -- Because of the small scale being used, analytical SFE is usually not concerned with these issues, while they are vital for successful process operations.

• **Cost** -- As previously mentioned, the initial capital cost for SFE can be high. Commercial analytical SFE systems cost $10,000 or more. However, this is generally not a major concern since the cost per extraction is much more favorable with carbon dioxide-based SFE compared with conventional organic solvent method. Concern for economics, however, is imperative for the success of process operations. In addition to the capital costs, energy costs are also an important consideration. Vijayan *et al.* (3) present three classifications of SFE process which may be economically favorable:

a) High-value, low-volume products such as the isolation of flavors, fragrances, and spices.
b) Intermediate-value, intermediate-volume products such as the decaffeination of coffee and tea and the deodorization of fats and oils.
c) Low-value, high-volume products such as the processing of oilseeds. This is perhaps the area where the processing economics must be particularly scrutinized.

The following chapters present a variety of applications of supercritical fluids to natural products. SFE, especially using carbon dioxide, is particularly well-suited to the extraction of natural products, especially essential oils and lipophilic materials. The use of SFE for natural products applications can be found in tomes by Stahl, Quirin, and Gerard (4) and Rizvi (ed.) (5) and in several literature reviews (6-11). Table II provides an overview of these applications. Based on experience and a survey of the literature, several guidelines for extraction with supercritical carbon dioxide have been suggested (4). These include:

• Lipophilic compounds (including hydrocarbons, esters, ethers, ketones, and related materials) with molecular mass up to 300-400 are easily extractable at pressures up to 5000 psi.
• Increasing solute polarity decreases solubility.
• Polar substances like sugars, glucosides, amino acids, lecithins, and polymers are not extractable, though nonpolar oligomers can be extracted.
• Water is slightly soluble with solubility increasing with temperature.
• Fractionation is possible when the compounds of interest display differences in molecular mass, vapor pressure, or polarity.

Based on physical chemical considerations, the future for SFE applications appears attractive. SFE will find a home in cases where the low temperature, selectivity, and high mass transfer rate advantages can be exploited. Especially on the process-scale, considerations with phase behavior and cost must be addressed. However, most successful process operations involving SFE are concerned with biologically produced materials and polymers, while analytical SFE has focused on environmental and natural products applications. Consequently, SFE is uniquely suited to address both analytical and processing issues with natural products.

Table II.
List of Selected Materials Extracted with Supercritical Fluids

Carotenoids
  β-Carotene
  Lycopene
  Lutein

Flavor and Fragrance Compounds
  Terpenoids
  Essential Oils
  Hops Bittering Acids
  Sesquiterpenes
  Monoterpenes
  Diterpenes
  Aliphatic Hydrocarbons
  Sesquiterpene Lactones
  Oxygenated Sesquiterpenes
  Furanocoumarins
  Vanillin
  Gingerol

Lipids, Fats, and Oils
  Fatty Acids
  Fatty Acid Esters
  Phospholipids
  Tocopherols
  Monoglycerides
  Diglycerides
  Triglycerides
  Prostaglandins

Mycotoxins
  Aflatoxins B1
  Fumonisin B1
  Trichothecenes

Alkaloids
  Caffeine
  Nicotine
  Pyrrolizidines
  Thebaine
  Codeine
  Morphine
  Isoquinolines
  Protopine
  Allocryptopine
  Phenanthridones
  Indoles
  Cocaine
  Quinine

Triterpenes and Sterols
  Cholesterol
  Stigmasterol
  Testosterone
  Cortisone
  Ergosterol
  Estriol
  Diosgenin

Other
  Pheromones
  Taxanes
  Allylbenzenes
  Cuticular Waxes
  Pyrethrins

Literature Cited

(1) Bartle, K. D.; Clifford, A. A.; Hawthorne, S. B.; Langenfeld, J. J.; Miller, D. J.; Robinson, R. J. *Supercrit. Fluid* **1990**, *3*, 143-149.
(2) Hawthorne, S. B.; Galy, A. B.; Schmitt, V. O.; and Miller, D. J. *Anal. Chem.* **1995**, *67*, 2723-2732.
(3) Vijayan, S.; Byskal, D. P.; Buckley, L. P. In *Supercritical Fluid Processing of Food and Biomaterials;* Rizvi, S. S. H., Ed.; Blackie Academic & Professional: New York, NY, 1994; pp 75-92.
(4) Stahl, E.; Quirin, K.-W.; Gerard, D. *Dense Gases for Extraction and Refining;* Springer-Verlag: New York, NY, 1986.
(5) *Supercritical Fluid Processing of Food and Biomaterials;* Rizvi, S. S. H., Ed.; Blackie Academic & Professional: New York, NY, 1994.
(6) Smith, R. D. *LC-GC* **1995**, *13*, 930-939.
(7) Modey, W. K.; Mulholland, D. A.; Raynor, M. W. *Phytochem. Anal.* **1996**, *7*, 1-15.
(8) King, J. W. *J. Chromatogr. Sci.* **1990**, *28*, 9-14.
(9) Castioni, P.; Christen, P.; Veuthey, J. L. *Analusis* **1995**, *23*, 95-106.
(10) Randolph, T. W. *Trend Biotechnol.* **1990**, *8*, 78-82.
(11) Williams, D. F. *Chem. Eng. Sci.* **1981**, *36*, 1769-1788.

Chapter 6

Extraction of Sage and Coriander Seed Using Near-Critical Carbon Dioxide

O. J. Catchpole[1], J. B. Grey[1], and B. M. Smallfield[2]

[1]Industrial Research Limited, P.O. Box 31–310, Lower Hutt, New Zealand
[2]Crop and Food Research Limited, Private Bag 50034, Mosgiel, New Zealand

Extraction of dried sage and coriander seed was carried out using near critical carbon dioxide to obtain oleoresin, non-volatile oil and essential oil extracts. Extractions were carried out in both a 4 litre and 75 litre extraction plant, to determine the effects of the particle size, carbon dioxide flow rate, bulk density and extraction temperature and pressure on the yield and extraction time. The rate of extraction of oleoresin from sage depended only on the particle diameter, and was limited by intra-particle diffusion. The rate of extraction of non-volatile oil from coriander seed was limited both by its solubility in carbon dioxide, and intra-particle diffusion. The results were satisfactorily correlated with a mathematical model. Scale-up calculations were performed to enable the economics of the extraction process to be evaluated.

Near critical extraction is assuming increasing importance in the food processing arena. Large scale plants have been in operation for a number of years that produce decaffeinated coffee and tea, and hop extracts for beer manufacture (*1-5*). Near-critical extraction of a wide range of herbs and spices is also carried out at a commercial scale in Western Europe and Japan (*1,6*) for producing essential oils and food flavourings. Industrial scale herb and spice extraction plants are operated in a semi-batchwise manner with respect to the solids. The solids are contained in a number of extraction vessels which are exposed to continuous recirculation of carbon dioxide (*5,6*). Each vessel is exposed for a certain extraction time to the recirculating fluid before being depressurized for removal of the spent solids and eventual recharge.

To obtain the best economic performance from such a plant, the production rate of extract must be maximised. This entails maximising the yield per batch or vessel load of solids, and minimising the extraction time. The number of batches extracted per unit time is directly related to the extraction time and number of extraction vessels. The yield per batch depends on properties of the solid and how well the bed is packed. Uneven packing can lead to axial mixing, channelling, and excessive pressure drop. The maximum theoretical yield per batch is determined by the concentration of extract in the feed material, feed moisture content, bed bulk density and particle size. The rate of extraction usually has both kinetic and equilibrium constraints. The kinetic constraints are a function of the particle diameter and flow rate. The equilibrium constraints are a function of the solubility of the extract in the fluid and the equilibrium relationship between the concentration of the extract in the fluid and in the solid. Other factors that influence the yield and chemical composition are the genotype of herb or spice used, and the season in which harvesting took place.

In this work, the effects of particle size; carbon dioxide flow rate; method of bed packing; and extraction temperature and pressure on the yield and extraction time for sage and coriander seed are reported. Sage is a typical herb which contains essential oil, waxes and water as the only CO_2 extractable components. Coriander seed is a typical spice which contains a non-volatile fraction (triglycerides), essential oil, and water that is extractable using near-critical carbon dioxide. The results are used for determining the technical and economic feasibility of near-critical herb and spice extraction.

Experimental

Flowering Dalmatian sage (*Salvia Officinalis*) and coriander seed (*Coriandum sativum*) were supplied by Crop and Food Research Limited, Mosgiel, New Zealand. The sage was supplied in a dried state. The sage included leaves, stems and flowers. The particle size of sage and coriander seed was reduced using a knife mill. The maximum particle size was determined by a sieve plate with holes of a known size which was fixed under the rotating blades. The chopped herb or spice was placed in a basket which had a solid wall and porous plates at the top and bottom. The basket and solids were then weighed, and placed in the extraction vessel of either a pilot (3.75 litre) or demonstration scale (75 litre) plant. A schematic of the demonstration scale plant is shown in Figure 1.

To begin an extraction, the apparatus was pressurized and heated to the desired temperature and pressure. When the pressure and temperature had been reached, carbon dioxide was then passed downwards through the packed bed, and then through a depressurization valve. The carbon dioxide and extract then entered the separation vessel where boiling and superheating took place, which caused precipitation of the extract. Vapour phase carbon dioxide was then either re-compressed by a compressor (pilot scale); or a liquid pump (demonstration scale) after condensation and sub-cooling. Further heat exchange was required to bring the temperature to the extraction temperature before passing through the extraction vessel again. The pilot scale extraction plant, described in detail elsewhere (7), had two separation stages so that fractionation of extract was possible. The first was operated at a pressure of 95 bar and temperature of 316 K to precipitate triglycerides. The second separator (the only separator on the demonstration scale plant) was operated at a pressure of 50-65 bar and over a temperature range of 323-328 K.

Sage extractions were carried using liquid carbon dioxide, at a pressure of 70 bar and temperature between 285 and 293 K. Coriander seed extractions were carried out at pressures in the range 70-320 bar and temperatures of 298-333 K. The

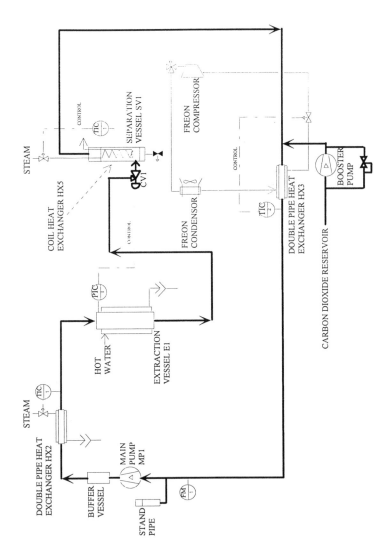

Figure 1. Schematic of demonstration scale (75 litre) extraction plant

temperature was controlled to $\pm$ 0.5°C, and the pressure to $\pm$ 0.5 bar. Most of the two stage separation experiments using coriander seed were performed at 250 bar and 313 K. The extract from the last separator was discharged into a settling vessel to separate into oil and water phases by gravity. Three particle sizes were used for both the sage and coriander extractions. The particle size was estimated by using sieves to separate a known initial mass into fractions retained according to their particle size range, and then calculating the weight mean particle diameter from the mass retained per sieve.

Three methods were used to pack the extraction baskets for the sage experiments. These were designated gravity fill; stick compression; and plunger compression. Gravity fill simply involved pouring the chopped sage into the basket in discrete intervals, followed by tapping the outside of the basket walls to cause settling. Stick compression also involved the discrete addition of sage to the packed bed, but was then followed by compression with a cylindrical rod. Plunger compression was a variation of the stick compression method, where a plunger with the diameter of the bed was used to compress the whole cross-section. Plunger compression was also used for pilot scale coriander experiments, while the gravity fill technique was used at the demonstration scale. Compression resulted in packed bed bulk density increases of up to 25 % for sage, but only 2.5 % for coriander. The bed bulk densities as a function of the particle size and packing method are listed in Table I for both the sage and coriander seed experiments. The packed height in the extraction baskets for pilot and demonstration scale experiments was 0.28 and 1.97 m respectively, with usable basket volumes of 2.96 litres and 65 litres respectively.

Table I.		Packed bed bulk densities (in kg m^{-3})		
material	particle diameter, mm	plunger	bulk density gravity	stick
sage	0.68	337.9	253.4	368.4
	0.85	304.1	226.4	294.0
	0.98	261.9	179.1	250.1
coriander	0.59	473.1		
	0.90	473.1	461.5	
	1.52	371.7		

Mathematical Model

A mathematical model was required for scale-up calculations that could take into account the effects of the process parameters temperature, pressure, particle diameter, bed bulk density and carbon dioxide flow rate at different scales of operation. Suitable models are available from the adsorption literature, as extraction from a packed bed of solids is analagous to desorption from a packed bed of solid adsorbent particles that are initially pre-saturated with solute. The following assumptions were made to enable an analytical model to be obtained: no axial or radial dispersion; uniform initial concentration of oil in the solids q_0; a constant interstitial solvent velocity U; a linear adsorption isotherm q = KC; and a parabolic concentration gradient in the particle (linear driving force approximation). The mass balance equations are analagous to those for the saturation of adsorbent particles *(8,9)*, and are thus not repeated here. Manipulation of the mass balance equations according to the assumptions above gives rise to a simple partial differential equation that relates the fluid phase concentration C to the dimensionless bed height x, and extraction time θ *(7-9)*:

$$\frac{\partial C}{\partial x} = -\xi \frac{\partial^2 C}{\partial \theta \partial x} - K \frac{\partial C}{\partial \theta} \tag{1}$$

$$\xi = R\left\{\frac{K}{3k_f} + \frac{R}{15D_e}\right\} \tag{2}$$

where ξ is a combined mass transfer resistance. An analytical solution of this equation is possible, and has been presented for many different initial conditions *(8-12)*. For the case under consideration here, the carbon dioxide entering the bed is assumed to be free of solute at all times $[C(0,\theta) = 0]$, and the initial concentration of solute in the particles is uniform at time $\theta = 0$ $[q_R(x, 0) = q_{av}(x, 0) = q_0]$. The solution to equation 1 subject to these initial conditions is as follows:

$$\frac{C}{C_0} = 1 - J(Kx/\xi, \theta/\xi) \tag{3}$$

$$\frac{q}{q_0} = J(\theta/\xi, Kx/\xi) \tag{4}$$

The J function has been tabulated *(13)*. An approximate method for calculating the J function *(10)* has been used in this work for z ≥ 1, where z = $2(uv)^{0.5}$, and for equation 3 u = Kx/ξ , v = θ/ξ. The approximation has been given for the Thomas function $\Phi(u,v)$ which is related to the J function by equation 6, with the J function given as equation 5.

$$J(u, v) = 1 - \exp(-v) \int_0^u \exp(-\beta) I_0[2(v\beta)^{1/2}] d\beta \tag{5}$$

$$1 - J(u, v) = \exp(-u-v) \phi(u, v) \tag{6}$$

The yield of extract, E, as a percentage of the total extract available is given by equation 7:

$$E=\frac{100}{KX}\int_0^\theta \frac{C}{C_0}\,d\theta \qquad (7)$$

For z < 1, a two term approximation for I_0 (z) can be used in equation 5, to give the following expression for C/C_0:

$$\frac{C}{C_0}=\exp(-v)\{1+v-\exp(-u)\,[1+v+uv]\} \qquad (8)$$

Equation 8 holds at very small times θ, or when both the rate of extraction is intra-particle diffusion controlled, and the solubility of the extract in carbon dioxide is very high compared to the initial concentration of the extract in the herb material. Mathematically, this is expressed as $K/3k_f \ll R/15D_e$ and $K \ll 1$. Substitution of equation 8 into equation 7, followed by integration and simplification gives the following simple model equation:

$$E=100\,(1-\exp[-15D_e\theta/R^2]) \qquad (9)$$

The general model (equation 3) parameters include a film mass transfer coefficient k_f, an effective diffusivity D_e, and an equilibrium solubility C_0. The film mass transfer coefficient can be calculated from a dimensionless correlation given elsewhere *(14,15)*, the effective diffusivity is obtained by optimising the fit of the model to the experimental data, and the equilibrium solubility C_0 for triglyceride oils can be obtained from a literature correlation *(16)*. Under the extraction conditions investigated in this work, the solubility of essential oil is very high or completely miscible *(17)* so that $K \ll 1$. The solubility is not required when equation 9 is applicable, as it does not appear in the final equation.

Results and Discussion

Sage. All sage extractions were carried out using liquid carbon dioxide. The sage raw material contained stems, leaf and flower heads. The extracts obtained were a mixture of water, waxes and essential oil. Water was easily separated from the wax and essential oil phase by using a separating funnel. The wax/essential oil phase was a brown liquid at room temperature, with a strong characteristic aroma of sage. The maximum yield of this liquid was 20g/kg (2 % by weight) of dried sage. When leaf only was extracted, the yield increased to 28 g/kg. This value approaches the yield for non-flowering sage reported elsewhere *(7)*. Supercritical carbon dioxide extraction of sage essential oil and waxes has also been recently reported *(18)*. A yield of 1.35 % by weight for essential oil was obtained, after separating the waxes out in a second separation stage *(18)*. The particle diameter was found to have the greatest influence on the rate of extraction and yield of essential oil per batch of solids extracted. As the particle size increased, the extraction time to reach maximum yield increased sharply. The flow rate had very little influence on either yield or extraction time. The yield of essential oil as a function of extraction time is shown in Figure 2 for three carbon dioxide flow rates and two packing methods. A model curve using an average yield

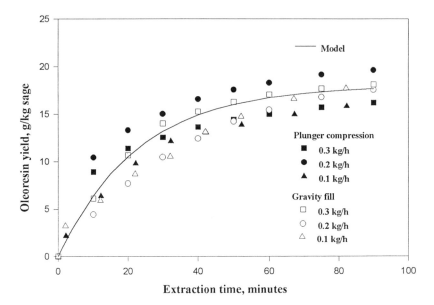

Figure 2. Yield versus extraction time for three flowrates and two packing methods and a particle diameter of 0.68 mm

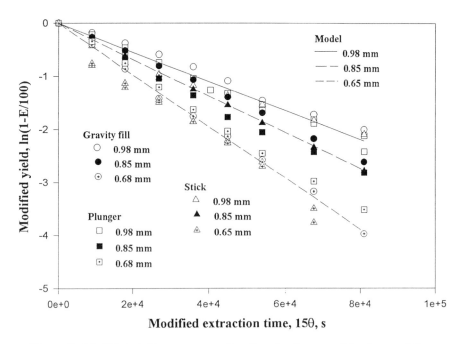

Figure 3. Modified yield versus extraction time for three particle sizes and three packing methods

of 18 g/kg of sage is shown for comparison. Although there is considerable scatter as the sage feedstock was not homogeneous, the two packing methods and different flow rates do not give rise to variations in the extraction time, or yield per kilogram of sage. These findings are consistent with the rate of extraction being controlled by intra-particle diffusion. Extraction curves for a wide range of other herbs reported in the literature including peppermint (*19*), basil, rosmary, marjoram (*20*), parsley, fennel (*21*), and most recently, sage *(22)* show a similar profile to Figure 2 All extraction results could be fitted by the simplified model equation 9. Equation 9 can be rearranged to obtain the effective diffusivity D_e from the experimental data:

$$\ln(1-E/100) = -15 D_e t / R^2 \tag{10}$$

A plot of the modified yield versus the extraction time is shown for the three particle sizes in Figure 3, for all three packing methods and a constant CO_2 flow rate of 0.2 kg/min. Again, while there is considerable scatter due to variation in the yield, the results for the three particle sizes clearly fall on separate lines. The effective diffusivity was found to be 6.12×10^{-12} m^2 s^{-1}. Numerical solution of the mass balance equations after making assumptions equivalent to that of a parabolic concentration profile in the solids, resulted in an effective diffusivity of 6.0×10^{-13} m^2 s^{-1} for the extraction of sage essential oil *(22)*. Compression of the bed resulted in identical yields per unit mass of solids, and thus a higher yield per experiment at a pilot scale, but lower yields per unit mass at a demonstration scale. This was attributed to channeling of fluid in the lower sections of the bed. Compression of the bed caused difficulties in removing the solids after completion of extraction, especially at the larger scale. Bed compaction is thus not recommended for obtaining higher yields per batch of solids for long packed beds and large scale operation. Since a lowering of the yield was not observed in the pilot scale experiments with relatively short bed lengths, it may be possible to use bed compaction at a larger scale if the bed is divided into sections which are separately packed.

Coriander.   Coriander seed extraction was carried out using supercritical carbon dioxide with pressures in the range 200-320 bar at a temperature of 313 K, and also liquid carbon dioxide at 70 bar and 293 K. A faintly yellow coloured essential oil extract with an aroma characteristic of freshly crushed herb was obtained at a maximum yield of 1.5 % by weight from 1.6 kg of coriander seed when using two stage separation. The yield of essential oil was similar to that reported by Moyler (*5*). The extraction behaviour of the essential oil was very similar to the oleoresin from sage, and has been reported elsewhere (*7*). The rate of extraction could be increased by decreasing the particle diameter, but not by increasing the flow rate. The rate of extraction of essential oil seemed to be independent of that for triglycerides, which was very dependent on flow rate. At a fixed particle diameter, the same yield of essential oil was obtained as a function of time irrespective of the yield of triglyceride. Again, the extraction kinetics for the essential oil are similar to those from other spices including cardamom, clove, cumin, and black pepper (*21,23*).

The triglyceride oil was obtained from coriander seeds at a maximum yield of 75 g/kg seeds. The oil was initially light yellow, but darkened considerably as the completion of extraction was approached. The oil was liquid at room temperature, and

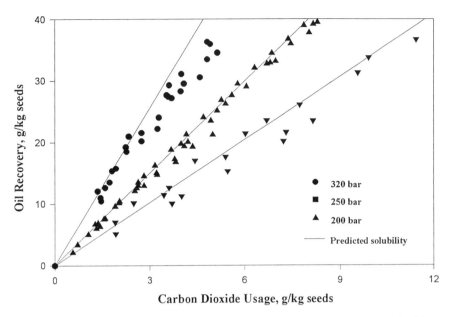

Figure 4. Extraction yield during constant rate period versus carbon dioxide usage for three pressures at 313 K

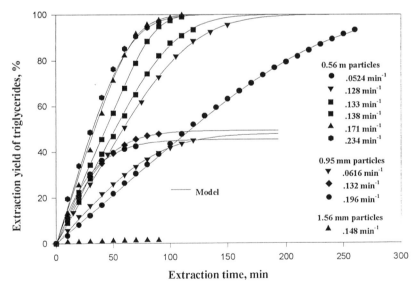

Figure 5. Percentage of theoretical oil yield versus extraction time for three particle sizes and a range of specific flow rates

solid at refrigeration temperatures. The carbon dioxide flow rate, particle diameter and extraction temperature and pressure were all significant in determining the rate of extraction and total yield of triglycerides from coriander seeds. The rate of extraction was initially limited by the solubility of the oil in carbon dioxide. In Figure 4, the oil yield as a function of carbon dioxide usage (cumulative carbon dioxide recirculated through the packed bed) over the initial constant rate extraction period is shown for three extraction pressures at 313 K. Also included is the predicted solubility from the correlation of Del Valle and Aguilera (*16*). There is good agreement between the predicted solubility and the cumulative oil yield. A constant rate extraction period for triglycerides has also been observed when extracting other seeds with significant quantities of triglycerides such as rape seed (*24,25*), tomato seed (*26*), and evening primrose seed (*27,28*). The percentage of the maximum triglyceride yield as a function of extraction time is shown for a variety of flow rates and three particle sizes in Figure 5. The effect of flow rate and particle size on yield can be clearly seen. Increasing the particle size resulted in a reduced yield of triglycerides. At the largest particle diameter (seeds broken into halves), very small amounts were recovered. Increasing the flow rate resulted in a reduction in the extraction time. The cumulative yields were well correlated by the model.

Extraction Economics.

The economic performance of a general purpose herb and spice extraction plant with vessel volumes from 10 to 300 litres was estimated, based on the sage and coriander extraction results. A liquid carbon dioxide plant was costed for the extraction of herbs (sage), and a supercritical extraction plant for spices (coriander). Extraction economics for a general purpose herb and spice extraction plant have been estimated for extraction plant with vessel volumes in the range 1000-4000 litres (*29*). This scale is too large for New Zealand conditions as insufficient crop is grown. The optimum number of 500 litre extraction vessels for a general purpose spice extraction plant using black pepper as a model spice, which recovers both a non-volatile extract (flavour components) and volatile extract (essential oil), was found to be three (*6*). Three extraction vessels have thus been assumed for this work. The operating and capital costs have been calculated on trading conditions in New Zealand at the time of publication, and converted to $US. The following utility costs have been assumed: carbon dioxide at $0.5/kg, electricity at $0.1/kW h, steam at $20/tonne, and cooling water at $1/tonne. Utility usage was scaled from that measured during 75 litre demonstration scale experimental runs.

The investment cost and operating cost per tonne of solids extracted per annum for both the liquid carbon dioxide and supercritical carbon dioxide extraction plants are shown in Figure 6 as a function of the extraction vessel capacity. There is a definite economy of scale, as both investment and operating costs per tonne of material extracted decrease with increasing plant capacity. A Net Present Value (NPV) calculation was performed to give the minimum selling price (MSP) of extract to achieve a 15 % annual return on capital invested. The NPV of an investment is the present value of the net cash flows out of the investment less the present value of the cash (capital) investment (*30*). The annual rate of return was chosen to be significantly higher than that available from banks and financial institutions of around 5-7 % per

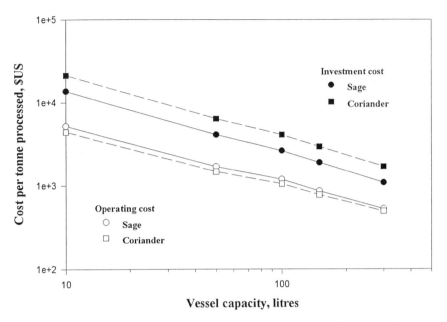

Figure 6. Investment and operating costs per tonne of feed processed as a function of vessel volume

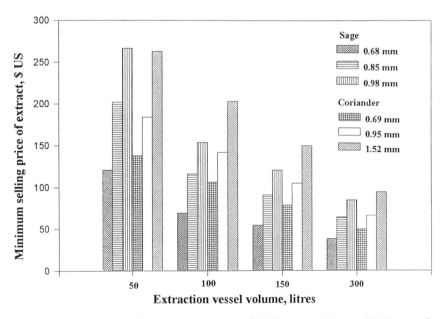

Figure 7. Minimum selling price of extract ($ US/kg) to achieve a 15 % rate of return on capital invested as a function of vessel capacity

annum. The MSP does not include the cost of the raw material. The following assumptions were used in the NPV calculations: maintenance costs are 2% of the capital cost, labour cost was $30,000.00 per worker, overheads are 50% of labour and maintenance costs, and the rate of depreciation was 20 % on the diminishing capital value of the plant. A 10 year plant life was also assumed. The MSP was estimated for 24 hour per day, 330 day per year operation. The MSP is shown for the three particle sizes at four plant capacities in Figure 7. Again, there is a favourable economy of scale, as the MSP decreases substantially as the plant capacity increases. The MSP also decreases as the particle diameter decreases, due to the decrease in extraction time, and improvement in packed bed bulk density. Shorter extraction times mean that more batches of solids can be extracted per day. An increase in bulk density means that more extract is recovered per batch load of solids. Under the optimum conditions of a 300 litre extraction vessel volume and small particle diameter, the minimum selling price of sage is $ 38/kg and coriander $ 49/kg.

Conclusions

The extraction of sage and coriander using near critical carbon dioxide was carried out to establish the effects of particle size, bed bulk density, carbon dioxide flow rate and

Nomenclature

Symbol	Description	Units
C	fluid phase concentration at bed height Z	kg m^{-3}
C_0	fluid phase solubility	kg m^{-3}
C_{ps}	fluid phase concentration at particle surface	kg m^{-3}
D_e	effective or intra-particle diffusivity	m^2 s^{-1}
E	extraction yield, % of theoretical	
$I_0(z)$	modified Bessel function of zero order	
J(u,v)	J function defined by equation 5	
K	equilibrium coefficient = q_0/C_0	
k_f	film transfer coefficient	m s^{-1}
q	particle phase concentration	kg m^{-3}
q_0	initial particle phase concentration	
R,r	particle radius, radial parameter	m
t	time	s
U	interstitial velocity	m s^{-1}
u	J function variable, Kx/ξ	
v	J function variable, θ/ξ	
x	bed length parameter=$(1-\varepsilon)Z/\varepsilon U$	s
X	total dimensionless bed length= Lx/Z	
Z	bed length variable	m
z	J function variable = $2(uv)^{0.5}$	
β	dummy integration variable	
θ	extraction time = $t-Z\varepsilon/U$	
ξ	combined mass transfer resistance, equation 2	s
ε	bed voidage	

extraction temperature and pressure on yield and extraction time. The particle diameter was the most important parameter for determining the rate of extraction from sage, while the bed bulk density and flow rate were of lesser importance. The rate of extraction was limited by intra-particle diffusion. All parameters were important for determining the rate of extraction of triglycerides from coriander seed. The rate of extraction was limited by both film mass transfer and intra-particle diffusion resistances. The rate of extraction and yield from both materials was satisfactorially correlated by a mathematical model presented in this work. The economics of the extraction process were estimated for both sage and coriander as a function of plant capacity. There is a significant improvement in profitability with increasing scale of operation. The bulk density and particle diameter also have a significant influence on plant profitability. At the largest plant capacity, the minimum break even selling price for sage and coriander extract was $ 38 and $ 49 per kilogram respectively.

Literature Cited.

(1)   Krukonis, V.; Brunner, G.; Perrut, M. *Proceedings of Third Int. Symp. Supercritical Fluids,* Institut National Polytechnique de Lorraine, Strasbourg, **1994**, *vol 1*, 1.

(2)   Lack, E; Seidlitz, H. In *Extraction of Natural Products with Near-Critical Solvents*, King, M. B; Bott, T. R, (Eds); Blackie Acad. & Prof., Glasgow, **1993**; Chapter 5.

(3)   Gardner, D. S. In *Extraction of Natural Products with Near-Critical Solvents*, King, M. B; Bott, T. R, (Eds); Blackie Acad. & Prof., Glasgow, **1993;** Chapter 4.

(4)   Parkinson, G.; Johnson, E. *Chem. Eng.* July, **1989**, 36.

(5)   Moyler, D. In *Extraction of Natural Products with Near-Critical Solvents*, King, M. B; Bott, T. R, (Eds); Blackie Acad. & Prof, Glasgow, **1993**; Chapter 6.

(6)   Körner, J. P.; Bork, M. *Proceedings of the 3rd International Symposium on Supercritical Fluids*, Institut National Polytechnique de Lorraine, Strasbourg, **1994**, *vol2*, 229

(7)   Catchpole, O. J.; Grey, J. B.; Smallfield, B. M. *J. Supercritical Fluids*, in press, **1996**

(8)   Rice, R. G. *Chem. Eng. Sci.*,**1982**, 37,*1*, 83.

(9)   Liaw, C. H.; Wang, J. S. P.; Greenkorn, R. H.; Chao, K. C.. *AIChE. J.* **1979**, *25*, 376.

(10)   Thomas, H. C. *Ann. N. Y. Acad. Sci.* **1948**, *49*, 161.

(11)   Amundson, N. R. *J. Phys. Colloid Chem.* **1950**, *54*, 812.

(12)   Aris, R.; Amundson, N. R. *Mathematical Methods in Chemical Engineering Vol 2*, Prentice Hall, New Jersey. **1973**, pp 161-170.

(13)   *Perry's Chemical Engineers' Handbook*, Perry, R. H.; Green, D. Eds, McGraw-Hill, New York, **1984**, pp 16-28.

(14)   King, M. B.; Catchpole, O. J. In *Extraction of Natural Products with Near-Critical Solvents*, King, M. B; Bott, T. R. (Eds). Blackie Acad & Prof., London, **1993**; Chapter 7.

(15) Catchpole, O. J.; Simoes, P. J.; King, M. B.; Bott, T. R. *Proceedings of High Pressure Chemical Engineering, 2nd International Symposium*, Dechema-GVC, Erlangen, Germany, **1990**, 153.
(16) del Valle, J.; Aguilera, J. M. *Ind. Eng. Chem. Res.* **1988**, *27*, 1551.
(17) *Dense Gases for Extraction and Refining*, Stahl, E; Quiren, K.-W.; Gerard, D. (Eds), Springer-Verlag, Berlin, **1988**, Chapter 4
(18) Reverchon, E.; Taddeo, R.; Della Porta, G. *J Supercritical Fluids*, **1996**, *8*, 302
(19) Goto, M.; Sato, M.; Hirose, T. *J. Chem. Eng. Japan*, **1993**, 26, *4*, 401.
(20) Reverchon, E.; Donsi, G.; Sesti Osséo, L. *Ind. Eng. Chem. Res.* **1993**, *32*, 2721.
(21) Naik, S.; Lentz, H.; Maheshwari, R. C. *Fluid Phase Equilibria*, **1989**, *49*, 115
(22) Reverchon, E. *AIChE. J.* **1996**, *42*, 1763
(23) Meireles, M. A. A.; Nikolov, Z. L.. In *Spices, Herbs and Edible Fungi*, Charalambous, G., Ed., Elsevier, Amsterdam, **1994**, Chapter 6.
(24) Brunner, G. *Ber. Bunsenges. Phys. Chem.* **1984**, *88*, 887.
(25) Lee, A. K. K.; Bulley, N. R.; Fattori, M.; Meisen, A. *JAOCS*, **1986**, 63, *7*, 921.
(26) Roy, B. C.; Goto, M.; Navaro, O.; Hortacsu, O. *J. Chem. Eng. Japan*, **1994**, 27(6), 768.
(27) Catchpole, O. J.; Andrews, E. W.; Toikka, G. N.; Wilkinson, G. T. *Proceedings of Third Int. Symp. Supercritical Fluids*, Institut National Polytechnique de Lorraine, Strasbourg, France, **1994**, vol 2, 47.
(28) Lee, B-C.; Kim, J-D.; Hwang, K-Y.; Lee, Y. Y. In *Supercritical Fluid Processing of Food and Biomaterials*, Rizvi, S. S. H. (Ed). Blackie Acad. & Prof, London, **1994**, Chapter 13
(29) Novak, R. A.; Robey, R. J. In *Supercritical Fluid Science and Technology*, Johnston, K.P.; Penninger, J. M. L. (Eds). ACS Symp. Ser. 406, Washington DC, **1989**, Chapter 32
(30) Woinsky, S. G. *Chem. Eng. Progress*, **1996**, *3*, 33

Chapter 7

Supercritical Recovery of Eicosapentaenoic Acid and Docosahexaenoic Acid from Fish Oil

Christina Borch-Jensen[1], Ole Henriksen[2], and Jørgen Mollerup[1]

[1]Department of Chemical Engineering, Building 229, Technical
University of Denmark, 2800 Lyngby, Denmark
[2]FLS Miljø A/S, Environmental Management, Ramsingsvej 30,
2500 Valby, Denmark

Phase equilibrium measurements in systems of CO_2 and
fish oil fatty acid ethyl esters (FAEEs) at 70°C are
presented. Three different mixtures were investigated, a
natural FAEE mixture from the oil of the sand eel
(*Ammodytes sp.*), a urea fractionated mixture prepared
from the natural FAEE mixture, and a concentrate of $\omega 3$
FAEEs. The measurements were performed in a static type
apparatus making it possible to sample both the vapor and
liquid phases. The results presented include solubilities and
equilibrium ratios. The experimental results are used to
evaluate a process for the recovery of C20:5ω3 (EPA) and
C22:6ω3 (DHA) at 70°C.

Fractionation of fish oil fatty acid ethyl esters by supercritical CO_2 is a process that,
in our opinion, would have several advantages over traditional fractionation methods,
such as chromatography, crystallization, and molecular distillation. Using a supercritical
CO_2 process will reduce the amount of organic solvents to a minimum, thus minimizing
the residual solvents in the product, and because the process temperature is considerable
lowered compared to distillation, the risk of degradation of the product is minimized.
    Evaluation of a fractionation process using supercritical CO_2 as separation medium
requires knowledge of phase equilibrium data for the mixture to be fractionated. In this
work the experimental measurements include the mutual solubilities of FAEE-CO_2 and
the equilibrium ratios. The solubility data will indicate how much FAEE can be
dissolved in the CO_2 rich stream and how much CO_2 will be dissolved in the FAEE rich
stream, while the equilibrium ratios will indicate how the components are distributed.
Both types of information are essential when evaluating the possibilities of a
fractionation process. Literature on solubilities of FAEEs has been reviewed (1), but

only a few papers on natural mixtures of fish oil FAEEs and CO_2 exist (2-5). The equilibrium measurements are often followed by pilot scale experiments to establish the process conditions, number of trays, and the reflux ratio.

In this paper we present the results from the phase equilibrium measurements in CO_2 for three different FAEE mixtures: a natural FAEE mixture from the oil of the sand eel (*Ammodytes sp.*) (2), a urea fractionated mixture prepared from the natural FAEE mixture (3), and a concentrate of ω3 FAEEs. The urea fractionated mixture was prepared by crystallization of saturated and monounsaturated FAEEs with urea and ethanol as the solvent, followed by recovery of the polyunsaturates. The ω3 concentrate was prepared by chromatography. The purpose of measuring the data is to investigate the possibility of employing a supercritical fractionation process for the recovery of EPA and DHA. The measurements were performed on mixtures of different overall composition to investigate how the equilibrium ratios and the solubilities depend on the overall composition. The concentrations in mole % of selected components in the three mixtures are given in Table I.

Table I. Concentrations (mole%) of selected components from the three FAEE mixtures

FAEE	Natural	Urea fractionated	ω3 concentrate
C16:4ω3	0.9	3.5	0.2
C18:4ω3	3.8	13.8	2.0
C20:1ω9	3.8	0.6	-
C20:4ω3	0.6	1.0	2.8
C20:4ω6	-	0.9	-
C20:5ω3	9.3	35.8	51.0
C22:3ω3	-	-	1.7
C22:4ω3	-	-	1.1
C22:5ω3	0.4	1.1	2.5
C22:6ω3	8.3	30.7	38.4

Experimental setup and procedure.

The phase equilibrium apparatus was obtained from DB Robinson, Alberta Canada and has been described by Staby et al.(*1*). The apparatus consists of a pistoned phase equilibrium view cell in an air bath, a displacement pump, high pressure pycnometers, a rocking mechanism, filling, cleaning and sampling lines and a gasometer. The equilibrium cell is of the variable volume type where the desired pressure is established by a displacement pump. A schematic outline of the apparatus is shown in Figure 1.

Typically, the cell is filled with 15 cm^3 of FAEE feed mixture and 30 -80 cm^3 of CO_2. The measurements at the lowest pressures require the largest amount of CO_2 in the cell to ensure enough vapor phase to determine the solubilities. The cell is rocked for more than 3 hours to ensure proper mixing of the two phases. After equilibrium is achieved a 20 - 60 cm^3 sample of the CO_2 rich phase

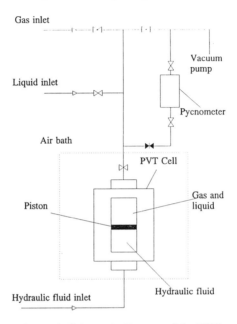

Figure 1. Schematic diagram of the PVT apparatus used for the experiments.

is withdrawn through a needle valve into a weighed and evacuated high pressure pycnometer. During sampling the pressure is maintained by displacing hydraulic fluid from the displacement pump into the cell. A 10-20 cm^3 sample of the FAEE rich phase is obtained in a similar manner after the sampling lines have been cleaned and evacuated.

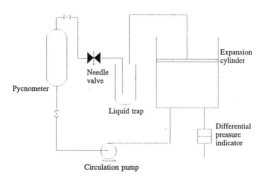

Figure 2. Schematic diagram of the gasometer

After sampling, the pycnometer is weighed again and the sample is degassed in the gasometer. The gasometer consists of a needle valve, a liquid trap and a 10 L expansion cylinder. Figure 2 shows a schematic outline of the gasometer. The pycnometer is connected to the gasometer and the CO_2 in the sample is carefully expanded to atmospheric pressure though the needle valve into the cylinder. The pressure difference between the inside of the cylinder and the atmosphere is less than 0.25 bar when degassing. The circulation system is used when large amounts of liquid are present in the pycnometer. After degassing the pycnometer is closed and weighed again. The solubilities are calculated from the measured masses of FAEE and CO_2 in each sample. After degassing the samples, the remains of the FAEEs from the vapor and liquid samples are poured out of the pycnometers and dissolved in n-Heptane to concentrations of 30 mg/mL and the samples are analyzed by gas chromatography. The equilibrium ratios on a CO_2 free basis are calculated from the chromatographic analysis.

Solubilities of fatty acid ethyl esters in CO_2.

The measured solubilities of the FAEE mixtures in carbon dioxide at 70°C are shown in Figure 3 as a function of the pure CO_2 density.

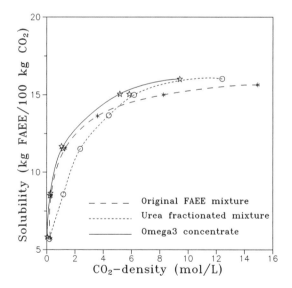

Figure 3. Solubilities of the three FAEE feed mixtures at 70°C as a function of the CO_2 density.

At low densities the solubilities increase dramatically with a small increase in the density while at higher densities the solubility curves are more flat. Generally, there are only small differences in the solubilities of the three different FAEE mixtures. This means that the change of mixture composition from a rather light, C16-rich, to a heavy, C20-22 rich, mixture does not have a dramatic effect on the solubility. In a separation process a large solubility is desired wherefore a high density should be selected. However, there is only little to gain by raising the CO_2 density from 7 to 14 mole/L, as can be seen from Figure 3. At 70°C, CO_2 densities of 7 and 14 mole/L correspond to pressures of 13 and 21 MPa respectively.

Equilibrium ratios of the FAEE components.
The equilibrium ratios, denoted by the K-values on a CO_2 free basis are obtained by analysis of the degassed vapor and liquid samples as shown in Figure 4 where K_i is the mass fraction of component i in the CO_2-rich vapor phase divided by the mass fraction of component i in the FAEE-rich liquid phase.

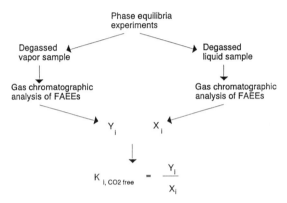

Figure 4. Calculation of K-values.

The K-values on a CO_2-free basis are related to the true equilibrium ratios through equation (1) where s_y and s_x are the mass of FAEE divided by the mass of carbon dioxide in the vapor and liquid phases, respectively. Doing the calculations on a CO_2

$$K_i = K_{i, CO_2-free} \frac{1 + 1/s_x}{1 + 1/s_y} \tag{1}$$

free basis makes the interpretation of the K-values more simple. If a component has a K-value below one it will be concentrated in the bottom product while components with K-value above one will be concentrated in the top product. The selectivity of two components, which is a measure for the possible separation, is the ratio of the K-values of the components.

Figures 5, 6 and 7 present K-values of four selected components from the three FAEE mixtures as a function of the system pressure at 70°C. All axes are identical which makes a direct comparison easy. The three figures illustrate the effect of changing the overall composition of the feed mixture. In Figure 5, which shows K-values for the natural FAEE mixture only one of the four components has a K-value above one and C18, C20 and C22 will thus go into the FAEE-rich bottom product, while C16 will be in the CO_2 rich top product. Using the urea fractionated FAEE mixture as the feed will affect the K-values as shown in Figure 6. Compared to Figure 5 it can be seen that in the urea fractionated mixture C18 is a light component with a K-value above 1. This is due to the fact that after the urea fractionation only the polyunsaturated FAEEs remain and because the saturated and monounsaturated components in fish oil FAEE happen to be the short chain FAEEs, the urea fractionation has made the original FAEE mixture more concentrated in polyunsaturates and of higher molecular weight. If the ω3

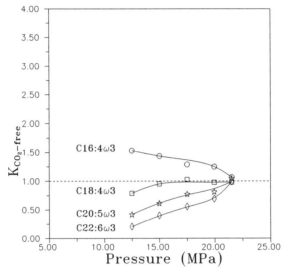

Figure 5. K-values of four selected components from the original FAEE feed mixture.

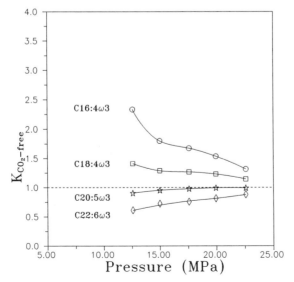

Figure 6. K-values of four selected components from the urea fractionated FAEE feed mixture.

concentrate is used as the feed mixture Figure 7 shows that both the C16, C18 and C20 components will be enriched in the light CO_2-rich top product which leaves C22 as the only heavy components. The Figures 5-7 thus illustrate how the separation is affected when some components are removed from the feed mixture which is what happens in a multi stage separation process. One characteristic feature in Figures 5-7 is that the K-values of the light components decrease with increasing pressure while the K-values of the heavy components increase with increasing pressure. This means that the selectivities decrease when the pressure is increased. The largest selectivities are thus obtained at low pressure. This means that the process pressure will be a trade off between solubility and selectivity. At 70°C a CO_2 density of 7 mole/L corresponding to a column pressure of 13 MPa, gives a satisfactory solubility and Figures 5-7 show that this pressure gives a sufficient selectivity between the C16, C18, C20, and C22 components.

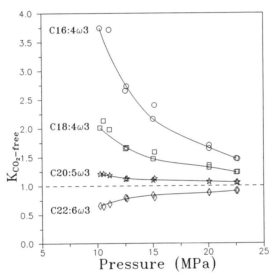

Figure 7. K-values of four selected components from the ω3-concentrate feed mixture.

Table I shows that EPA is not the only C20 component in the mixtures and DHA is not the only C22 component. In both cases isomers and different degrees of unsaturation exist and it is important to investigate whether a separation based on the degree of unsaturation is possible because this is essential when the scope is to obtain pure fractions of EPA and DHA. To illustrate this we present the K-values of the C20 and C22 FAEEs from the measurements with urea fractionated FAEEs and the ω3 concentrate. The results are shown in the Figures 8 and 9, respectively. The axes in these two Figures are identical to make a direct comparison easy. Figure 8 shows that the selectivities between the C20 isomers with the urea fractionated mixture are sufficient to separate components differing in the degree of unsaturation and also positional isomers. Figure 9 shows the K-values of the C20 and C22 components from the measurements of the ω3 concentrate. This Figure shows that the selectivities between the C20:4ω3 and C20:5ω3 has increased when compared to the urea fractionated mixture simply because of the change in the overall compositions of the mixtures. A

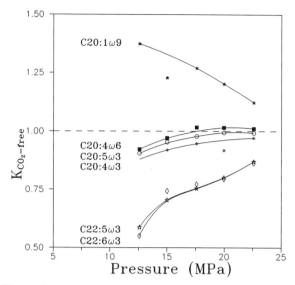

Figure 8. K-values of C20 and C22 components from the urea fractionated FAEE feed mixture.

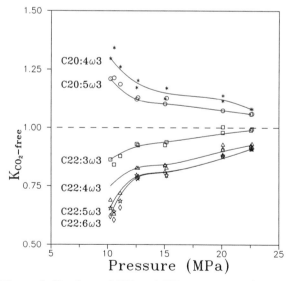

Figure 9. K-values of C20 and C22 components from the ω3 concentrate feed mixture.

fractionation to obtain a pure EPA fraction is thus possible. The selectivity between C22:5ω3 and C22:6ω3 (DHA) seems to be very small for both mixtures and the change in selectivities when changing the overall composition of the feed mixture seem neglible.

However the two remaining C20 components in the ω3 concentrate can be removed as a light extract, thus leaving the four remaining C22s in the bottom product. When this mixture is fed to the next stage of the process C22:3ω3 will become a light component compared to DHA and can then be removed from the top of the column and after that C22:4ω3 and C22:5ω3 is removed in the same manner. The purity of the DHA fraction will depend on the number of columns in the fractionation process and the number of plates in each column.

Fractionation of EPA and DHA from fish oil FAEes.

The results discussed in the previous section have shown the possibility of fractionation of EPA and DHA from fish oil FAEes in a fractionation process having several counter-current fractionation columns. Figure 10 shows the principle of such a column. The FAEes are fed at a given stage of the column while CO_2 enters the

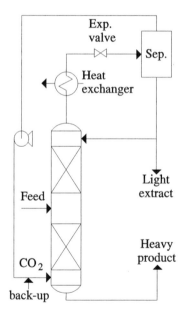

Figure 10. A supercritical fractionation process using a counter-current column.

column at the bottom. The flow direction of the liquid FAEes is downwards while the CO_2 flow is directed upwards in the column stripping off the light components in the feed. This process is called counter-current operation and is known to be much more efficient than batch operation. The heavy product leaves the column from the bottom while the light extract leaves the column from the top. The amount and composition of the extract is governed by the pressure and temperature of operation. After leaving the column, CO_2 can be recovered from the bottom and top product by decreasing the pressure or the product streams can be feed streams for other columns in the cascade. If one of the product streams is the final product, the released CO_2 is recycled after depressurizing. To provide a reflux, which is essential for the operation of the column, the extract stream leaving the top of the column is passed through a heat exchanger and a pressure regulator, the expansion valve. These devices are used to control the solvating properties of the CO_2 before the stream goes to the separator vessel where CO_2 and FAEEs are separated. After leaving the separator vessel part of the FAEEs are recycled back to the column in a reflux stream. The purpose of the reflux stream is to increase column efficiency. If the reflux is decreased, the required number of plates in the column will increase and vice versa. The CO_2 back-up stream compensates for any CO_2 lost during the process. When starting with a natural mixture of fish oil FAEEs, the number of different components can be as high as 50. In the case of a recovery process for EPA and DHA these two components are the only ones desired as the final products and the remaining components are considered as byproducts in this case. This means that it is important to design the fractionation process to remove the byproducts in as few steps as possible. Figure 11 shows a schematic outline of a fractionation process for EPA and DHA. The first step fractionates according to chain length to remove all the light C10 to C18 components from the mixture as a top product. The bottom product will be a C20 to C24 fraction. This fraction can now be urea fractionated to yield only the polyunsaturated C20 and C22 FAEEs. C24 components present in fish oil are saturated or monounsaturated and will be removed by the urea fractionation. The polyunsaturated C20 and C22 fractions can now be fractionated into a C20 and a C22 fraction from which EPA and DHA can be recovered by further fractionation. An alternative process scheme would be to urea fractionate the natural FAEE mixture before the first supercritical fractionation step. Which process to choose will depend on process economy. The first process will require a larger supercritical fractionation column in the first fractionation step and less urea than the latter.

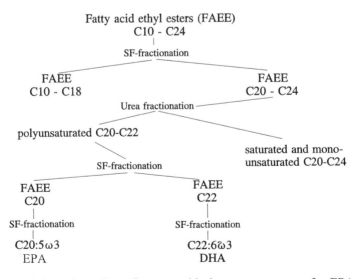

Figure 11. Schematic outline of a supercritical recovery process for EPA and DHA from fish oil FAEEs.

Conclusions.

In this work we have shown that a supercritical fractionation of EPA and DHA from fish oil FAEEs would be possible. We have presented the necessary phase equilibrium data to investigate the process scheme. The process evaluation has been based on solubilities and equilibrium ratios obtained from phase equilibrium measurements in our laboratory. The phase equilibria measurements predict what will happen in a single fractionation stage, wherefore pilot scale experiments are necessary to determine the number of stages in the column, the size of the column, the reflux and to optimize the operating pressure and temperature.

If urea fractionation is one of the intermediate steps in the process it would be beneficial to replace the organic solvents in the traditional urea fractionation process by supercritical CO_2. Urea fractionation of FAEEs is done in ethanol. This solvent is not particularly toxic, but a replacement of ethanol with CO_2 will reduce the complexity of the process because the other process steps are carried out using CO_2 as a solvent. Urea fractionation in supercritical CO_2 has been performed in pilot scale (6,7).

References.

(1)   Staby, A.; Mollerup, J., *Fluid phase equilibria*, **1993**, *91*, 349.

(2)   Staby, A.; Mollerup, J., *Fluid phase equilibria* , **1993**, *87*, 309.

(3)   Borch-Jensen, C.; Staby, A.; Mollerup, J., *Ind. Eng. Chem. res.*, **1994**, *33(6)*, 1574.

(4)   Krukonis, V., Vivian, J.E., Bambara, C.J., Nilsson, W.B., Martin, R.E; Paper presented at the *"194th ACS Meeting"*, New Orleans, **1987**.

(5)   Nilsson, W.B., Seaborn, G.T., Hudson, J.K.; *JAOCS*, **1992**, *69(4)*, 305.

(6)   Saito, S., *Kagaku Seibutsu*, **1986**, *24*, 201.

(7)   Suzuki, Y., *Kagaku Kogaku*, **1988**, *52*, 516.

Chapter 8

Supercritical Carbon Dioxide Extraction of Volatile Compounds from Rosemary

J. A. P. Coelho[1,4], R. L. Mendes[2], M. C. Provost[3], J. M. S. Cabral[1],
J. M. Novais[1], and A. M. F. Palavra[1]

[1]Departamento de Engenharia Quimica, Instituto Superior Tecnico, Complexo I,
Av. Rovisco Pais, 1096 Lisboa, Portugal
[2]Departamento Energias Renovaveis, Instituto Nacional de Engenharia
e Tecnologia Industrial, Estrada do Paço do Lumiar, Azinhaga dos Lameiros,
1699 Lisboa, Portugal
[3]U.S. Food and Drug Administration, Center for Devices and Radiological Health,
Rockville, MD 20850

Supercritical CO_2 extraction of oleoresin from portuguese rosemary
(*Rosmarinus Officinalis* L.) was carried out with a flow apparatus at
temperatures of 35 and 40 °C and pressures of 100, 125 and 200 bar.
The highest fraction of volatile compounds (oil) in the oleoresin was
obtained at 100 bar and 40 °C. An unsteady mathematical model was
able to give good representation of the supercritical extraction curves
and a mass transfer coefficient was determined using the successive
quadratic programming method. Values of this coefficient ranged
between 4.49 and 15.07 kg/m^3s and the shift to a diffusion-controlled
regime occurred when 44% of the total oil was extracted.

Rosemary is a strongly aromatic plant, whose essential oil has been used in the
perfume and food industries. Moreover, its oleoresin has antioxidant properties.
The conventional methods used to obtain oleoresin and essential oils present some
disadvantages. When extraction is performed with organic solvents, it is impossible to
remove all traces of the toxic solvent. The maximum acceptable residue level of some
solvents in foodstuffs and ingredients is strictly stated in the EEC directive (88/344)
and in the United States in Title 21 of the CFR published by the Food and Drugs
Administration (1). These documents impose great restrictions in solvents, such as
hexane methyl acetate and methylene chloride.
In the hydrodistillation, the high temperature and the presence of many monoterpene
hydrocarbons, which can be degraded through oxidation, lead to some instability in the
final product (2,3). These limitations can be avoided using supercritical fluid extraction
(SFE).

[4]Current address: Departamento De Engenharia Quimica, Instituto Superior de Egenharia de Lisboa,
Rua Conselheiro Emidio Navarro, 1,1900 Lisboa, Portugal

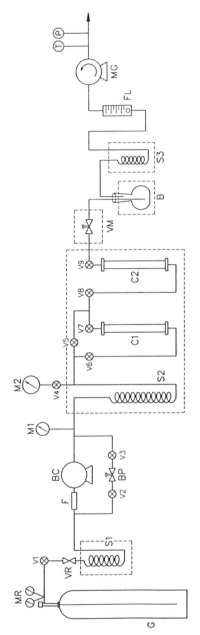

Figure 1 - Schematic diagram of the supercritical fluid extraction apparatus. G - cylinder, VR - check valve, S1-ice cooler, F - filter, BC - pump, BP - back-pressure regulator, M1,M2 - manometers, S2 - heat exchanger C1,C2 - extraction cells, VM - expansion valve, B - collection vessel, S3 - glass serpentine, FL - rotameter, MG - wet-test meter.

SFE is a separation technique in which the yield and selectivity can be controlled to some extent by changing the pressure and the temperature of the fluid. Carbon dioxide has been the most used supercritical solvent for application in the food and related industries. With this solvent it is possible to obtain free of toxic solvents extracts and to avoid the degradation of the thermal labile components. Therefore, the natural odor and flavor of the initial material are maintained.an unsteady model was applied by Lee *et al* (4) Schaffer *et al* (5) and Cygnarowicz-Provost *et al* (6) to the supercritical extraction of different compounds with CO_2.

The aim of this work was to study the extraction of oil from portuguese rosemary and to check if an unsteady model gives a good representation of the extraction curves.

Experimental

Apparatus
The diagram for the supercritical fluid extraction apparatus, described in detail in a previous paper (7), is presented in Figure 1. The separation section was modified for this work. Liquid CO_2 from a cylinder, G, was compressed to the desired pressure with the pump, BC. The pressure was controlled by a back-pressure regulator, BP, and measured with a Bourdon type manometer, M2 (± 0.5 bar). The carbon dioxide was preheated in a heat exchanger, S2, to reach the preset temperature before flowing into the extraction cells, C1, C2. The temperature was measured near the second cell with a platinum resistance thermometer ($\pm 0.1^{\circ}C$).

The supercritical fluid was expanded to atmospheric pressure through a micrometering valve, VM, into cold traps, a 250 cm^3 collection vessel, B, at -21 °C, and a glass serpentine, S3, at -71 °C. The flow rate was measured with a rotameter, FL, and the total volume of gas was determined with a wet test meter, MG (± 0.01 dm^3).

Material and Methods
In this work, 10 g of leaves of rosemary (the material was comminuted to a mean diameter, dp, of 1.33 and 0.72 mm) were submitted to the supercritical fluid at a flow rate of 0.4 L/min (STP). The amount of extract collected (oleoresin), in the cooled traps for a given time was determined gravimetrically with an analytical balance ($\pm$ 0.1 mg).

The volatile compounds (oil), in the extracts were analyzed by gas chromatography, GC. The analysis was performed using a Hewlett Packard 5890 gas chromatograph with a flame ionization detector. A fused silica column (50 m $\times$ 0.20 mm id.,film thickness 0.20 μm), coated with Carbowax 20M, was used. The oven temperature was programmed from 70 to 200 °C at 3 °C/min and kept at 200 °C for 30 min. The injector temperature was 250 °C and the carrier gas, He, had a flow rate of 0.5 mL/min.

To quantify the volatile compounds, cyclohexanol was added into the weighed extracts as an internal standard. The different responses of the compounds were taken into account by using calibration curves for the different families of compounds. Mass spectra were obtained with a Perkin Elmer 8320 chromatograph equipped with a Finnigan Mat ion trap detector (ITD-800). A fused silica (30 m $\times$ 0.25 mm, film thickness 0.20 μm) column, coated with Carbowax 20M, was used. Helium was the carrier gas.

Hydrodistillation was carried out in a Clevenger type apparatus and the essential oil obtained, 70.0 mg, was recovered by decantation.
The initial moisture content in the rosemary leaves was 9.6% by weight. After the supercritical extraction the moisture content ranged from 8 to 8.7%.

Extraction Modeling
In supercritical fluid extraction, SFE, the influence of pressure and temperature on solubility of different volatile compounds and oils has been described in the literature. However, the mass transfer mechanism of SFE of the natural products from the plant is not yet totally understood.
A mass transfer model is important to design supercritical extraction equipment. The unsteady model applied in this work was first used by Bulley *et al* (8) to describe the extraction curves of canola oil, assuming the existence of plug flow in the bed. The equations that describe the mass balance in the fluid and solid phase are, respectively:

$$\varepsilon\rho\frac{\partial y}{\partial t}+\rho U\frac{\partial y}{\partial h}=J(x,y)$$

$$(1-\varepsilon)\rho_s\frac{\partial x}{\partial t}=-J(x,y)$$

(1)

where ε is the bed porosity, ρ the solvent density, ρ_s the solid density, U the superficial velocity, t the time, h the distance from the bottom of the solid leaves, y the oil concentration in the fluid, x the mass of oil per mass of oil-free leaves and $J(x,y)$ the mass transfer rate.
The initial and boundary conditions are:

$$x = x_0 \qquad t = 0, \text{ any h}$$

$$y = 0 \qquad h = 0 \quad t \geq 0$$

(2)

To describe the mass transfer rate during the supercritical fluid extraction two different regimes can be considered . A first regime, when the accessibility to the oil is easier and only the diffusion resistance in the solvent controls the process. The second one, when the concentration of the oil in the plant decreases to a value below which the mass transfer is controlled by the diffusion in the solid phase.
Bulley *et al* (8) used a simple relation to describe the mass transfer rate in the first regime :

$$J(x,y) = A_p K(y^* - y)$$

(3)

where y^* is the equilibrium oil concentration and ApK the mass transfer coefficient. The equations 1 can be solved by the method of characteristics (9, 10).
As a constant mass transfer coefficient can not describe the extraction curve, Cygnarowicz-Provost (6) considered that the mass transfer coefficient decreased along

the extraction according to an empirical correlation which describe both regimes. A similar empirical correlation was applied, with success, by Mendes *et al* (11) to the SFE of lipids from algae:

$$J(x,y) = A_p K_0 \left(y^* - y\right) \exp\left[\ln(C) \frac{x_0 - x}{x_0 - x_d} \right] \qquad (4)$$

where x_d is the concentration of the oil in the plant at which the diffusion-controlled regime starts and C is a constant, determined by trial and error fits of the experimental data. In this work, the best representation of the curves was obtained when C=0.02. Note that in this expression, $ApK=ApK_0$ at the beginning of the extraction ($x = x_0$) and $ApK=0.02ApK_0$ when the shift to the diffusion-controlled regime occurs.
It was assumed that the maximum oil extractable by SFE, x_0, was the amount obtained by hydrodistillation and the equilibrium oil concentration, y^*, was determined by the slope of the extraction curve at the origin, considering that initially equilibrium was achieved. To compute the total oil extracted at any time, t, the following expression was used:

$$M_{oil} = Q \int_0^t y(t)dt \qquad (5)$$

where Q is the solvent mass flow rate.

Results and Discussion

Supercritical CO_2 extraction of oil from Portuguese rosemary was carried out at the temperatures of 35 and 40 °C and pressures of 100, 125 and 200 bar.
The effect of particle size on the amount of the oil extracted at 40 °C is shown in Figure 2. Reverchon (12) suggested that the oil is located in vacuoles inside the leaves of rosemary. When the particle size decreases, the access of the CO_2 to the oil is easier and therefore its concentration in the supercritical fluid increases at each pressure. Moreover, for each particle size, the amount of oil is higher at 200 bar.
For the rosemary with the smallest mean particle size, the highest fraction of volatile compounds (oil) in the oleoresin was obtained at 100 bar and 40 °C. Table I shows the content of these compounds obtained by hydrodistillation and supercritical extraction at these conditions of pressure and temperature.
The essential oil from hydrodistillation contains more monoterpene hydrocarbons which are unstable and their decomposition products deteriorate the quality of the flavor and the fragrance. On the other hand, the opposite trend was shown by the oxygenated monoterpenes.
Typical extraction curves for rosemary with the smallest mean particle size are shown in Figures 3 and 4 . Figure 3 shows the amount of essential oil obtained as a function of time at 35 and 40 °C and pressures of 100 and 125 bar. At the beginning of the extraction, the oil yield was similar for the two temperatures, while at the end it was higher at 40 °C. Figure 4   shows the amount of oil obtained at 40 °C and different

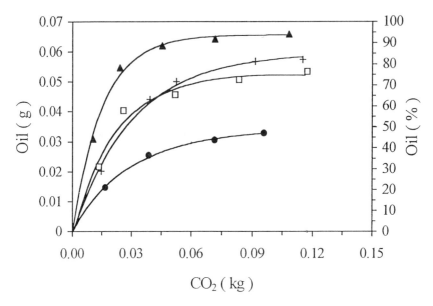

Figure 2 - Effect of particle size on the extraction of rosemary oil at 40 °C; dp=1.33 mm: ● 125 bar and + 200 bar; dp= 0.72 mm: □ 125 bar and ▲ 200 bar

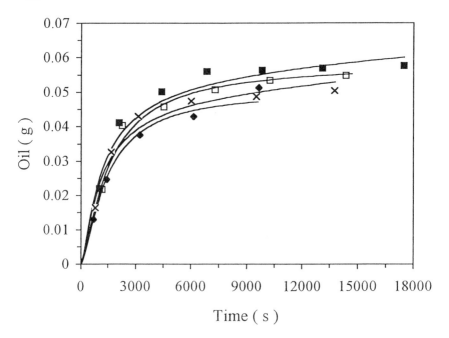

Figure 3 - Amount of rosemary oil as a function of time (Q=1.31x10^{-5} kg/s). dp=0.72mm: × 100 bar, 35 °C; ◆ 125 bar, 35 °C; ■ 100 bar, 40 °C; □ 125 bar, 40 °C; —— model.

Table I. Content of volatile compounds identified in rosemary oil (mg/g) obtained by hydrodistillation and supercritical extraction (dp = 0.72mm)

* Compound	Hydrodistillation	100 bar , 40 °C
α-Pinene	72.8	13.3
Camphene	29.2	8.3
β-Pinene	5.6	4.2
Sabinene	2.4	2.3
Myrcene	82.0	38.5
α-Phellandrene	3.4	3.0
α-Terpinene	3.1	3.0
Limonene	146.6	90.2
β-Phellandrene	4.6	1.6
1,8-Cineole	55.5	38.1
(Z)-β-ocimene	4.0	1.8
γ-Terpinene + (E)-β-ocimene	2.1	2.1
p-Cymene	27.3	19.6
Terpinolene	1.5	1.5
1-Octen-3-ol	2.7	3.1
Camphor	266.3	272.8
Linalool	8.7	10.1
Linaly Acetate	5.8	5.2
Bornyl Acetate	7.4	9.3
Terpinen-4-ol	7.1	10.4
β-Caryophyllene	15.3	23.0
α-Humulene	14.9	23.2
α-Terpineol	18.2	32.1
Borneol	15.0	21.8
Verbenone	19.5	70.6
β-Humulene	3.3	6.4
Caryophyllene oxide	5.5	11.6

*Compounds listed in elution order from the Carbowax20M column

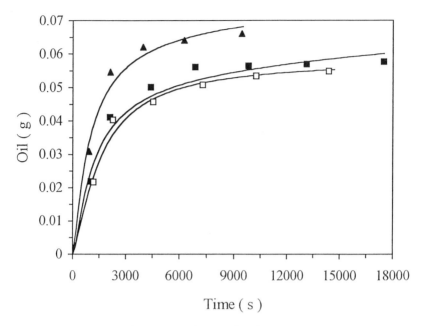

Figure 4 - Amount of rosemary oil as a function of time at temperature of 40 °C, (Q=1.31x10⁻⁵ kg/s). dp=0.72 mm: ■ 100 bar , ☐ 125 bar, ▲ 200 bar , —— model.

pressures. For 200 bar the amount of oil obtained at the end of the extraction was near the value achieved by hydrodistillation (70 mg).

The experimental data were correlated with the model, at different supercritical conditions, using an average seed density ρ_s, of 338.0 kg/m³, a bed porosity ε, of 0.5. The initial mass transfer coefficient, ApK_0, was determined (Table II).

Table II - Values for ApKo and y* at different pressures and temperatures (U=1.2x10⁻⁴ m/s)

P(bar)/T(°C)	dp(m)=7.2x10⁻⁴		dp(m)=13.3x10⁻⁴
	A_pK_0 (kg/m³s)	y^*(kg/kg)x10⁻³	A_pK_0 (kg/m³s)
100/35	4.49	1.88	
125/35	5.07	1.65	
100/40	7.76	1.98	
125/40	9.31	1.75	0.689
200/40	15.07	2.99	1.83

This coefficient increases with pressure at constant temperature and with temperature at constant pressure. For the particle size 7.2x10⁻⁴ m, the value at which the shift into the diffusion controlled regime occurred, x_d, was 3.67x10⁻³, while for the diameter

13.3×10^{-4} m was 3.95×10^{-3}. The found values to x_d reflect the efficiency of the milling process. Table II shows also, at constant pressure and temperature, that the smaller is the particle size larger is the mass transfer coefficient. The shift into the diffusion controlled regime occurred when about 48% and 44% of the total oil was extracted for the particle size 7.2×10^{-4} and 13.3×10^{-4} m respectively. For the rape seed oil, King *et al* (13) found a shift in the extraction when 65 % of initial content in oil was extracted. This difference could be due to the presence of the essential oil in the internal part of the vegetable matter (12, 14). On the other hand, the evaluated mass transfer coefficients in this work are on the same order of magnitude as those for other plants (15). Furthermore, a good representation of the extraction curves of volatile compounds with CO_2 from rosemary leaves was achieved using the unsteady state model.

References

1. Sanders, N., Extraction of Natural Products Using Near-Critical Solvents, Edits, King, M.B., Bott, T.R., Black Academic & Professional, Glasgow, U.K., **1993**, 35-37.
2. Reverchon, E., Senatore, F., Flav. Fragr. J., **1992**, 7, 227.
3. Reverchon, E., J. Super. Fluids, **1992**, 5, 256.
4. Lee, A.K.K., Bulley, N.R., Meisen, A., JAOCS, **1986**, 63(7), 921.
5. Schaeffer, S.T., Zalkow, L.H., Teja, A.S.J., J. Super. Fluids, **1989**, 2, 15.
6. Cyagnarowicz-Provost, M., O`Brien, D.J., Maxwell, R.J., Hampson, J.M., J. Super. Fluids, **1992**, 5, 4.
7. Mendes, R.L., Coelho, J.P., Fernandes, H.L., Marrucho, I.J., Cabral, J.M.S., Novais, J.M., Palavra, A.F., J. Chem. Tech. Biotechnol., **1995**, 62, 53.
8. Bulley, N.R., Fattori, M., Meisen, A., Moyls, L., JAOCS, **1984**, 61(8), 1362.
9. Acrivos, A., Ind. Eng. Chem., **956**, 48(4), 703.
10. Dranoff, J.S.; Lapidus, L. Ind. Eng. Chem., **1958**, 50(11), 1648.
11. Mendes, R.L., Fernandes, H.L., Cygnarowicz-Provost, M., Cabral, J.M.S., Novais, J.M., Palavra. A.F., Procee. of the 3rd Inter. Symp. Super. Fluids, Strasbourg, Fance, **1994**, Tome2, 477.
12. Reverchon, E., Dansi, G., Osséo, L.S., Ind. Eng. Chem. Res., **1993**, 32 (11), 2721.
13. King, M.B., Bott, T.R., Barr, M.J., Mahmud, R.S., Sanders,N., Sep.Sci.Technol., **1987**, 22, 1103.
14. Stahl, E., Gerard, D., Perfum. Flav., **1985**, 10(2), 29
15. Sovova, H., Procee. of the 3rd Inter. Symp. Super. Fluids, Strasbourg, France, **1994**, Tome2, 131.

Chapter 9

Effect of Functional Groups on the Solubilities of Coumarin Derivatives in Supercritical Carbon Dioxide

Y. H. Choi[1], J. Kim[1], M. J. Noh[2], E. S. Choi[2], and K.-P. Yoo[2]

[1]College of Pharmacy, Seoul National University, 151–742, Seoul, Korea
[2]Department of Chemical Engineering, Sogang University, C.P.O. Box 1142, Seoul, Korea

Solubilities of the basic coumarin, five mono-substituted derivatives (7-hydroxy-, 7-methoxy-, 7-methyl-, 6-methyl- and 4-hydroxy-coumarin) and three di-substituted derivatives (6,7-dihydroxycoumarin, 7-hydroxy coumarin-4-acetic acid, and 7-methoxycoumarin-4-acetic acid) in supercritical CO_2 were measured in a range of temperatures 35~50 °C and pressures 8.5~25 MPa. In general, mono-substituted coumarin derivatives were less soluble than the basic coumarin in CO_2. The degree of solubility tends to increase in the order of 7-methyl-, 7-methoxy-, 7-hydroxy-, and 4-hydroxycoumarin. However, in the case of 6-methylcoumarin, it was unusually more soluble than the basic coumarin by several times in CO_2 over the entire experimental conditions. Furthermore, the di-substituted coumarin derivatives were extremely less soluble than the basic coumarin and monosubstituted coumarins. For each coumarin derivative, optimum equilibrium conditions which gives maximum solubility in CO_2 were reported.

It has been experimentally proven that supercritical CO_2 can be an efficient alternative solvent over the conventional organic liquid solvents for many cases of selective extraction of bioactive substances from natural plants (*1-4*). The supercritical fluid extraction (SFE) is more economical than liquid solvent extraction (*5-8*). It is the usual cases that the total extracts from natural plants with any means are supposed to contain many compounds and their numerous substituted derivatives. Thereby, whenever one intends to obtain a highly purified target substance from the total extract it is essential to know at least the quantitative solubility behavior of each solute in a specific supercritical solvent. The information of phase equilibrium behavior between a solute and a solvent are imperative for the preliminary evaluation of an economic viability of the SFE on process for a given sample material.

110

Thus, many investigators have been reported the phase equilibrium data of natural substances in supercritical fluids (*9-11*). However, such reliable thermodynamic data are extremely scarce in most cases.

Take for instance, reports on the quantitatively compared solubility for a specific natural substance together with its derivatives are extremely limited. Thus, in the present study, we placed our attention on the solubility measurement of coumarin and its derivatives over a wide range of supercritical conditions.

Since coumarin and its substituted derivatives have been known for their strong bioactivities such as antibiotic, antitumor, vasodilatory and anticoagulant (*12*), some researchers have been concerned with the extractability of these substances from plants with supercritical CO$_2$ (*13-15*). However, no systematic and quantitative solubility studies are demonstrated yet for the purpose of reliable process design. Thus, in the present study, we placed our attention on a comparative measurement of several coumarins in supercritical CO$_2$.

Among numerous types of coumarins, we illustratively selected nine types of coumarins with emphasis on the fixed positions (e.g., C-4 and C-7 position as shown in Figure 1) and tried to measure their respective solubilities. Tested derivatives include coumarin derivatives substituted by with methyl-, methoxy-, hydroxy-, and acetic acid group, respectively.

Experimental

Chemicals.   Reagent-grade (<98% purity) coumarin and substituted coumarins (7-hydroxy-, 4-hydroxy-, 6-methyl-, 7-methoxy-, 7-methyl-, 6,7-dihydroxy-, 7-hydroxy coumarin-4-acetic acid and 7-methoxycoumarin-4-acetic acid) were purchased from Sigma and Aldrich Co. (Milwaukee, WI, USA) and used them without further purification. CO$_2$ (< 99.95%) was obtained from Seoul Gas Co. (Seoul, Korea). Other solvents of HPLC-grade used in chromatographic analysis were purchased from J. T. Baker. (ST. Louis, MO, USA).

Solubility Measurement.   A schematic diagram of the experimental flow-type micro-scale apparatus used in the present study is shown in Figure 2. Volume of the equilibrium cell is 60 mL. Pressure was controlled by a gas booster (HASKEL75/15, Burbank, CA, USA) and Heise gauge (HEISE MM-43776, Stratford, Connecticut, USA). Temperature in the air-bath was controlled by PID controller (Hanyoung, Seoul, Korea) within ±1 ℃. The effluent coumarin solutes dissolved in supercritical CO$_2$ were collected by a 2-step methanol trap and the extract that remained in the equipment lines and valves were rinsed three times after every experimental runs by methanol, acetone and chloroform, respectively. Through the repetitive preliminary tests, the most stable state of the effluent flow rate was controlled at 200 ml/min at ambient condition (25 ℃, 1 atm) by a metering valve (HIP 60-11-HEV-V, Erie, Pennsylvania, USA). A mass flowmeter (Sierra 8810, Chicago, IL, USA) was used to measure the amount of CO$_2$ consumed. For each sample solute, experiments were repeated five times to ensure the reliable reproducibility and accuracy of the measured solubility data. Each solubility of solutes was measured at three isotherms (35, 40 and

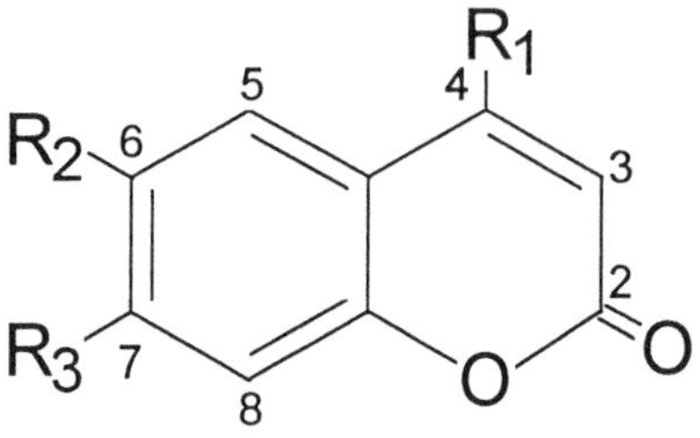

$R_1 = R_2 = R_3 = H$ coumarin

$R_1 = OH, R_2 = R_3 = H$ 4-hydroxy coumarin

$R_1 = R_2 = H, R_3 = OH$ 7-hydroxy coumarin

$R_1 = H, R_2 = R_3 = OH$ 6,7-dihydroxy coumarin

$R_1 = R_2 = H, R_3 = OCH_3$ 7-methoxy coumarin

$R_1 = R_2 = H, R_3 = CH_3$ 7-methyl coumarin

$R_1 = R_3 = H, R_2 = CH_3$ 6-methyl coumarin

$R_1 = CH_2COOH, R_2 = H, R_3 = OH$ 7-hydroxy coumarin-4-acetic acid

$R_1 = CH_2COOH, R_2 = H, R_3 = OCH_3$ 7-methoxy coumarin-4-acetic acid

Figure 1. Coumarin and the derivatives measured their solubilities in supercritical CO_2 in the present study.

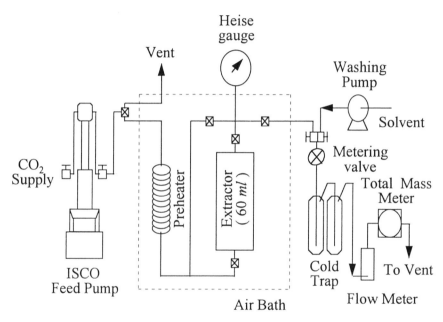

Figure 2. Schematic diagram of micro-scale SFE Apparatus.

50 ℃) and at each isotherm, pressure was varied by 8.5, 10, 15, 20 and 25 MPa, respectively.

HPLC Determination of Solubility of Coumarins. All the coumarins chosen in this study have their own chromophore and it can be easily identified by UV detector. Thereby, we used HPLC instead of gas chromatography. Each extract collected during each solubility measurement was dried in a vacuum evaporator in order to be securely free from the solvents in the cold traps and from the rinsing steps. As a next step, each dried sample is redissolved by methanol and analyzed by a HPLC (Milton Roy, Riviera Beach, FL, USA).

The HPLC is equipped with Constametric 3000 pump, Spectromonitor 3100 variable wave length detector fixed at 280 nm and Chromocorder 12 (SIC System Instruments CO, Tokyo, Japan). The column was YMC-Pack, ODS-A (250 × 4.6 mm, s-5 μm) (YMC Inc. Kyoto, Japan). The mobile phase used in this experiment was a mixture of 50 by 50 ratio of methanol and 5% formic acid (*16*). For each solute, calibration curves were prepared and they were used as a reference for assaying each sample. Finally, to check whether the coumarins are thermally decomposed or not during experimentation, we analyzed the solute before and after experiment by HPLC and conformed the thermal stability within the entire range of experimental temperatures.

Results and Discussion

Measured solubilities of nine coumarins expressed by mole fraction (y_2) in supercritical CO_2 were shown in Table I. In general the solubility of monosubstituted coumarins except the 6-methylcoumarin tends to increase with increasing pressure. At low pressure (8.5 MPa), the solubility tends to decrease with an increase in temperature for all compounds as summarized in Table I. At high pressure (25 MPa) the solubility tends to increase with an increase in temperature for all compounds with the exception of the basic coumarin and 6-methylcoumarin. At the intermediate pressures the trends are less clear and depend on the specific compound and condition.

In Figure 3, the solubility variation of basic coumarin with different equilibrium conditions were shown. As one can see, the maximum solubility is detected at 40 ℃. To see an effect of different types of groups in a fixed position (C-7 position as shown in Figure 1) on the solubility in supercritical CO_2 , solubilities of coumarin, 7-methyl-, 7-methoxy-, and 7-hydroxycoumarins were measured in the range of 35~40 ℃ and 8.5~25 MPa as shown in Figure 4. Substitution of one functional group tends to decrease the solubilities in the order of methyl, methoxy and hydroxy groups. Also, to examine an effect of a same group in the different position on the solubility, the solubilities of 4-hydroxy- and 7-hydroxy-coumarin were compared to each other. As a result, we found that 7-hydroxycoumarin dissolves about twice time more than the case of 4-hydroxycoumarin in supercritical CO_2 as shown in Figure 5. In these experiments, we found that the existence of hydroxy group in the basic coumarin structure tends to decrease significantly their solubilities than the case of the basic coumarin.

To see the effect of further addition of hydroxy group on the solubility of

Table 1. Solubilities (y_2) of coumarin and its derivatives in supercritical on dioxide at various temperature, pressure

$T/^\circ C$	P/MPa	coumarin $y_2 \times 10^3$	4-hydroxy-coumarin $y_2 \times 10^6$	7-hydroxy-coumarin $y_2 \times 10^6$	7-methoxy-coumarin $y_2 \times 10^4$	7-methyl-coumarin $y_2 \times 10^3$	6-methyl-coumarin $y_2 \times 10^3$
35	8.5	0.996(±0.035)	0.093(±0.008)	0.223(±0.012)	0.165(±0.058)	0.368(±0.042)	1.167(±0.011)
	10.0	1.183(±0.117)	0.168(±0.028)	0.358(±0.013)	0.880(±0.269)	0.691(±0.129)	3.097(±0.096)
	15.0	1.430(±0.196)	0.337(±0.032)	0.677(±0.096)	1.288(±0.019)	1.131(±0.069)	7.814(±0.217)
	20.0	1.790(±0.194)	0.411(±0.069)	1.029(±0.213)	2.946(±0.794)	1.217(±0.074)	11.68(±0.389)
	25.0	2.119(±0.137)	0.454(±0.042)	1.472(±0.239)	4.099(±0.024)	1.396(±0.159)	10.65(±0.706)
40	8.5	0.452(±0.112)	0.022(±0.008)	0.033(±0.019)	0.051(±0.011)	0.103(±0.020)	0.745(±0.139)
	10.0	1.532(±0.007)	0.128(±0.035)	0.366(±0.028)	0.557(±0.085)	0.656(±0.211)	2.597(±0.605)
	15.0	1.683(±0.193)	0.324(±0.006)	0.808(±0.065)	1.631(±0.212)	1.317(±0.069)	6.300(±0.455)
	20.0	1.873(±0.288)	0.487(±0.066)	1.254(±0.054)	3.091(±0.182)	1.606(±0.041)	8.109(±0.249)
	25.0	2.659(±0.284)	0.598(±0.017)	1.517(±0.034)	4.004(±0.618)	1.793(±0.066)	10.65(±0.615)
50	8.5	0.062(±0.009)	0.012(±0.007)	0.015(±0.003)	0.005(±0.0003)	0.033(±0.012)	0.600(±0.036)
	10.0	0.231(±0.061)	0.079(±0.002)	0.134(±0.036)	0.457(±0.056)	0.163(±0.034)	1.291(±0.015)
	15.0	0.457(±0.064)	0.413(±0.034)	1.032(±0.210)	2.092(±0.035)	1.219(±0.046)	8.343(±0.733)
	20.0	0.756(±0.048)	0.877(±0.023)	1.659(±0.079)	3.931(±1.610)	1.719(±0.048)	15.38(±1.370)
	25.0	0.902(±0.075)	1.021(±0.007)	2.270(±0.095)	4.546(±0.098)	2.165(±0.089)	5.061(±0.370)

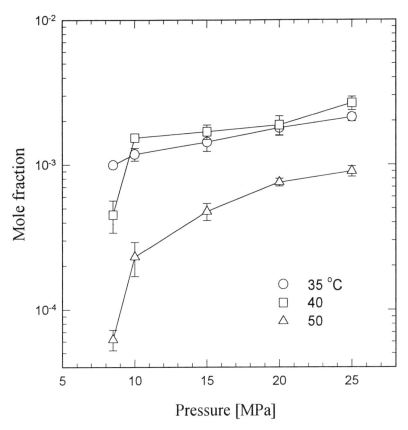

Figure 3. Measured solubilities of basic coumarin in supercritical CO_2 in the ranges of 35~50 ℃ and 8.5~25 MPa.

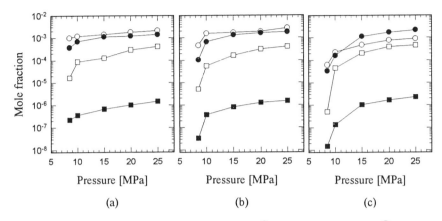

Figure 4. Compared solubilities of coumarin (○), 7-methylcoumarin (●), 7-methoxycoumarin (□), and 7-hydroxycoumarin (■) in supercritical CO_2 at 35 ℃ (a), 40 ℃ (b) and 50 ℃ (c).

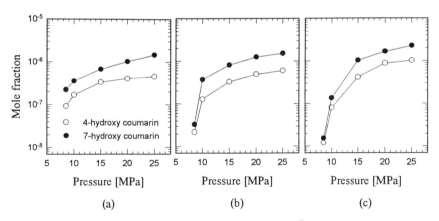

Figure 5. Compared solubilities of 4-hydroxycoumarin (○), 7-hydroxycoumarin (●), in supercritical CO_2 at 35 ℃ (a), 40 ℃ (b) and 50 ℃ (c).

coumarin in supercritical CO_2, the solubility of 6,7-dihydroxycoumarin was measured and we found that this derivative showed almost no solubility at all (y_2 were lower than $10^{-10} \sim 10^{-12}$). In the case of 6,7-dihydroxycoumarin, even the addition of tracer amount of cosolvent such as methanol could not cause it to dissolve into supercritical CO_2 as reported by others (*13,14*).

Also, we found that the addition of acetic acid group (i.e., 7-hydroxycoumarin-4-acetic acid and 7-methoxycoumarin-4-acetic acid) drastically reduce their solubilities in supercritical CO_2.

In summary we found that addition of methyl group into the basic coumarin (6-methyl-, 7-methyl-) tends to increase the solubility in supercritical CO_2. However, the addition of other functional groups (i.e., hydroxy-, methoxy-, and acetic acid) tends to reduce drastically their solubilities in supercritical CO_2. Further quantitative experimental study is in progress by the present authors for the effect of solubility variations of coumarin derivatives with respect to their additional position versus different functional groups in supercritical fluids.

Acknowledgments

The authors are grateful to the Korea Science and Engineering Foundation for financial support. They also express their thanks to the Sunkyung Industry, Korea for a grant.

Literature Cited

1. Bevan, C. D.; Marshall, P. S. *Nat. Prod. Rep.* **1994**, *11*, 451.
2. Stahl, E.; Quirin, K. W.; Gerard, D. *Dense Gases for Extraction and Refining*; Springer-Verlag: New York, 1988, pp 72-217.
3. Vilegas, J. H. Y.; Lancas, F. M.; Vilegas, W.; Pozzeti, G. L. *Phytochem. Anal.*1993, *4*, 230.
4. Xie, Q. L.; Markides, K. E.; Lee, M. L. *J. Chromatogr. Sci.* **1989**, *27*, 3 65.
5. Smith, R. M.; Burford, M. ,D. *J. Chromatogr.* **1992, *627*, 250.
6. Liu, B; Lockwood, G. B; Gifford, L. A. *J. Chromatogr. A.* **1995**, *690*, 250.
7. Bicchi, C.; Rubiolo, P.; Frattini, C.; Sandra , P.; David, F. *J. Nat. Prod.* **1991**, *54*, 941.
8. Choi, Y. H.; Kim, J.; Noh, M. J.; Park, E. M.; Yoo, Y. -P. *KJChE.*1996, *13*, 216
9. Chen, J. W.; Tsai, F. N.; *Fluid Phase Equilib.* **1995**, *107*, 189.
10. Tan, C. S.; Weng, J. Y. *Fluid Phase Equilib.* **1987**, *34*, 37.
11. Krukonis, V. J.; Kurnik, R. T. *J. Chem. Eng. Data,* **1985**, *30*, 247.
12. Murray, R. D. H.; Mendez, J.; Brown, S. A. *The Natural Coumarins;* John Willey & Sons: New York, 1982, pp13-20.
13. Calvey, E. M.; Page, S. W. *J. Supercrit. Fluids,* **1990**, *3*, 115.
14. Miachi, H.; Manabe, A.; Tokumori, T.; Sumida, Y.; Yoshida, T.; Nishibe, S.; Agata, I.; Nomura, T.; Okuda, T. *Yakugaku Zasshi* **1987**, *107*, 435.
15. Miachi, H.; Manabe, A.; Tokumori, T.; Sumida, Y.; Yoshida, T.; Kozawa, W.; Okuda, T. *Yakugaku Zasshi* **1987**, *107*, 367.
16. Casteele, K. V.; Geiger. H.; Sumere, C. F. *J. Chromotogr.* **1983**, *258*, 111.

Chapter 10

Supercritical Fluid Extraction with Reflux for Citrus Oil Processing

Masaki Sato[1], Motonobu Goto[2], Akio Kodama[2], and Tsutomu Hirose[2]

[1]Department of Industrial Science, Graduate School of Science and Technology, and [2]Department of Applied Chemistry, Kumamoto University, 2–39–1 Kurokami, Kumamoto 860, Japan

Terpenes in citrus oil must be removed to stabilize the products and to dissolve it in aqueous solution. Supercritical fluid extraction has been investigated for the terpeneless citrus oil processing as a lower temperature process. In order to achieve higher yield and higher separation selectivity, a continuous countercurrent extraction with reflux was studied at a temperature of 333 K and a pressure of 8.8 MPa. Cold-pressed orange oil from Brazil and a model mixture of 80 % limonene and 20 % linalool, where limonene and linalool are principal constituent of terpenes and oxygenated compounds in orange oil, were used as feed and carbon dioxide was used as solvent. Operation at total reflux was carried out to calculate the minimum number of plates required to achieve a separation between limonene and linalool. Effects of the solvent-to-feed ratio, reflux ratio, and feed inlet position on the yield and selectivity were investigated for continuous operation. The selectivity increased with the increase in the solvent-to-feed ratio. Terpeneless citrus oil was obtained on the operation at the higher solvent-to-feed ratio and longer stripping section.

Citrus oil processing is important in the perfume industry. Cold-pressed citrus oil is a mixture of high volatile components such as terpenes and oxygenated compounds, and non-volatiles such as pigments and waxes. Terpenes in citrus oil must be removed to stabilize the product and dissolve it in aqueous solution. They are conventionally processed by distillation or solvent extraction, resulting in thermal degradation of the compounds. Supercritical fluid extraction with carbon dioxide as solvent has been focused for the terpeneless citrus oil processing as an alternative process. This process makes it possible to operate at lower temperature than conventional processes (1-2).

However, simple extraction with supercritical carbon dioxide has not been successful since it is difficult to optimize the operating conditions in terms of both the extraction yield and separation selectivity. Namely, a lower pressure led to a higher selectivity but a lower extraction yield, whereas a higher pressure gave a higher extraction yield but a lower selectivity. Extensive research to overcome this problem has been carried out in the last decade. Yamauchi and Saito (3), Barth et al. (4), and Chouchi et al. (5) introduced adsorbent into a supercritical fluid system to achieve higher selectivity. They successfully fractionated lemon oil, bergamot oil, and orange oil by using silica gel as an adsorbent. However, since the desired oxygenated compounds are strongly adsorbed, a considerably high pressure is required to desorb the solutes, resulting in a requirement for higher energy input and higher instrument costs.

In the extraction process, more effective separation can be established by increasing the number of the equilibrium stages for a continuous countercurrent extraction as well as multi-stage operation. Perre et al. (6) developed the countercurrent extraction process, that successfully obtained terpeneless oil concentrated 5 or 10 fold with an operation of 0.25 m^3 of feed oil per day on the ECOSS pilot plant (125 mm i.d. x 8 m height). On the other hand, Gerard (7) suggested that internal reflux induced by a temperature gradient was advantageous in the enriching section for the production of terpeneless essential oil by supercritical carbon dioxide.

High pressure phase equilibrium data are needed for the design of the supercritical fluid extraction process. Matos et al. (8) measured phase equilibria for the system of limonene-CO_2 and cineole-CO_2, at a temperature of 318 - 323 K. Giacomo et al. (9) measured the solubility of limonene and citral in CO_2 at 308 - 323 K. More recently, Shibuya et al. (10) and Suzuki and Nagahama (11) reported the solubility of limonene and linalool in CO_2 at a temperature of 313 - 333 K for orange oil processing. Furthermore, Iwai et al. (12) measured phase equilibria for the system of linalool - CO_2 at 313 - 333 K and correlated those data with the Peng-Robinson equation of state (PR-EOS) (13).

Unfortunately, available equilibrium data are limited to binary systems. Equilibrium data for the ternary system are vital for the design of extraction process. Cubic equations of state such as PR-EOS and SRK-EOS are helpful for estimating phase equilibria. However, these equations include binary interaction parameters, which are usually obtained by fitting with experimental data. Previously, we studied the semi-batch extraction of a mixture of limonene and linalool with supercritical carbon dioxide, and obtained the equilibrium data of vapor phase for the ternary system CO_2 - limonene - linalool (14). Good correlation with experimental data by PR-EOS was obtained at 333 K and 8.8 MPa.

In this work we evaluate the performance of the continuous countercurrent extraction with reflux for the removal of terpenes in citrus oil. Operation at total reflux, including an internal and an external reflux, was carried out for understanding the limiting operating conditions. This may give information concerning the minimum number of stages required to achieve a separation when no feed enters and no product is withdrawn. In a continuous operation, we investigated the effects of the solvent-to-feed ratio, external reflux ratio, and feed inlet position on the separation selectivity and the yield.

Experimental

Materials. Cold pressed orange oil from Brazil (supplied from Givaudan-Roure Flv. Ltd.) and a model mixture of 80 wt% limonene (Kanto Chem. Ltd.) and 20 wt% linalool (Kanto Chem. Ltd.) were used as feed. Liquefied carbon dioxide was obtained in a cylinder with siphon attachment (Uchimura Sanso Ltd.) and was used as an extraction solvent.

Apparatus. A schematic diagram of the bench scale experimental apparatus is illustrated in Figure 1. The column is 20 mm internal diameter and the total column

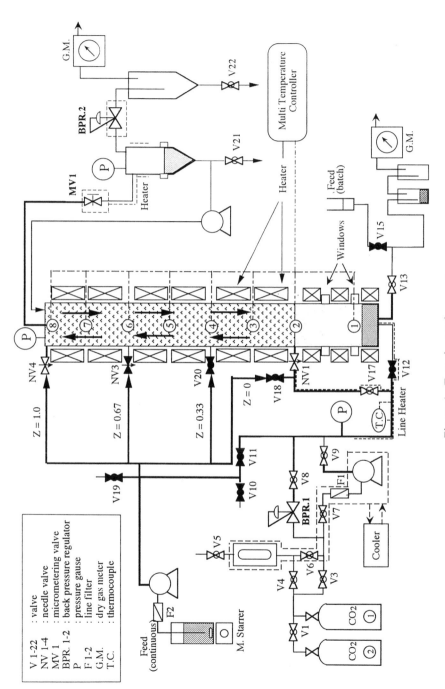

Figure 1 Experimental apparatus

length is 2400 mm. The rectification part is packed with stainless steel 3 mm Dixon Packing in a length of 1800 mm. Two pairs of 10 mm diameter windows are located at the bottom of the column to permit visual observation of the interface between liquid and vapor. The temperature of the column is controlled by eight PID controllers.

The separator for external reflux is 40 mm i.d. and the total volume is 600 ml. The fluid exiting from the column is cyclonically introduced to the separator. One pair of 10 mm diameter windows is also located at the bottom of the separator. The separator was kept at a pressure of 5 MPa and a temperature of 303 K.

Carbon dioxide was passed through a cooled line and compressed to operating pressure by a high pressure pump. The pressure was controlled by a back pressure regulator (BPR.1). The compressed CO_2 was passed through a line heater in order to reach the desired temperature. The fluid exiting from the top of the column was expanded to ambient pressure through a micrometering valve which was heated by a ribbon heater to prevent it from freezing. The CO_2 flow rate was adjusted by the micrometering valve (MV1).

Operation at total reflux. For total reflux operation, 300 g of feed was charged into the bottom of the column through valve 13 up to the level of the higher window and into the bottom of the separator up to the window level. CO_2 bubbled through valve 12 into the feed to start the extraction. External reflux was controlled by a high pressure pump and the interface between liquid and vapor in the separator. Sampling was carried out at certain intervals from both the separator and the column. About 30 hours were required to reach steady state.

The CO_2 flow rate used ranged from 0.1 - 0.5 g/s. The extraction pressure was 8.8 MPa, and the temperature of the column was kept at 333 K. The effect of internal reflux in the column induced by a temperature gradient from 313 K at the bottom of the column to 333 K at the top was investigated in comparison with that of the uniform temperature at 333 K.

Continuous operation. The column was also operated in the countercurrent flow mode, where CO_2 was the continuous phase entering through the needle valve (NV1) and feed oil was the dispersed phase entering at a certain inlet position of the column. In this procedure, the height of the interface between vapor and liquid phase did not affect the selectivity. Feed inlet position, Z, indicates the length of the stripping section divided by the length of the total rectification column of 1800 mm. Upper and lower sections of the column act as enriching and stripping regions, respectively. A feed inlet position was set by valve operation. Flow rate of the feed oil was kept constant at 0.422 g/min. The extract and the raffinate were periodically withdrawn from the bottom of the separator and from the column, respectively. External reflux ratio was controlled by the visual adjustment of the interface through the window, the flow rate of the high pressure pump, and the amount of extract. In this work, it took more than 5 hours to achieve steady state.

Analysis. The extract and raffinate were collected in sampling tubes in a certain interval of time, weighed, and analyzed by a capillary gas chromatograph. The GC analysis was performed with a Shimadzu GC-14A instrument equipped with a J&W DB-WAX (30 m x 0.25 mm i.d., 0.25 mm film) fused-silica capillary column and FID. The temperature of the GC oven was programmed linearly from 343 K to 473 K at a rate of 3 K/min. Helium was used as carrier gas with split ratio of 1/100.

Phase equilibrium

Data for the ternary system (1) CO_2 - (2) limonene - (3) linalool are not available in the literature. Phase equilibria for this system were therefore calculated by PR-EOS with binary interaction parameters. The binary interaction parameters were obtained from optimization with our previous experimental data (14). The data of vapor phase for the ternary system were obtained from the experiment with semi-batch extraction, where the extract leaving the extractor was assumed to be in equilibrium with the liquid in the extractor (14). In the optimization, the binary interaction parameter, k_{13}, which is the binary parameter between CO_2 and linalool, is adopted from the value of binary system (12).

Figure 2 shows the phase equilibrium data on both a Janecke diagram and a X-Y diagram on a CO_2-free basis at 333 K and 8.8 MPa. Phase equilibrium data for binary system measured by Shibuya et al. (10), Iwai et al. (12)and Suzuki et al. (11) are also shown in Figure 2. Calculated phase equilibrium has the binary interaction parameters, $k_{12} = 0.274$, $k_{13} = 0.051$, and $k_{23} = -0.026$, where the subscript numbers, 1, 2, and 3, correspond to CO_2, limonene, and linalool, respectively. Calculated line with PR-EOS was in good agreement with experimental data. The average values of the separation selectivity obtained from the experimental results and from the calculated phase equilibrium are 2.05 and 1.95, respectively. The separation selectivity was defined by the following equation.

$$\beta = \frac{Y'_2 / X'_2}{Y'_3 / X'_3}$$

where the prime indicates a CO_2-free basis. The selectivity on the equilibrium corresponds to that of the simple extraction.

Results and Discussion

Operation at total reflux. When all the exiting solutes from the column return to the column, there is no product and the number of stages is a minimum for any given separation. Under this condition, the operating line lay simply on the diagonal line in the X-Y diagram shown in Figure 2.

Figure 3 shows the change of limonene concentration (CO_2-free) in extract and raffinate as a function of CO_2 flow rate. When the column has a temperature gradient from 313 K at the bottom to 333 K at the top, not only an external reflux but also an internal reflux may occur in the column. The rectification due to the internal reflux will be useful for the batch operation (14-15) or the enriching section in a continuous operation (7). Furthermore, the temperature gradient from 313 K at the bottom to 333 K at the top at a pressure of 8.8 MPa indicated highest selectivity (14). However, in the citrus oil process for the total reflux operation, the raffinate could not be withdrawn for sampling at the bottom of the column on this operation, because the homogeneous phase is formed at the bottom of the column at 8.8 MPa and 313 K. Namely, the solutes are dissolved in supercritical CO_2 completely and the liquid phase is not formed. Therefore, a temperature gradient over this temperature range is not suitable for the continuous operation of citrus oil processing.

On the other hand, for the operation using the column with an uniform temperature of 333 K, limonene concentration was dramatically reduced with an increase in flow rate of CO_2. In this case only the external reflux was given. Thus, a higher flow rate of CO_2 favored the separation of citrus oil. The separation selectivity between limonene and linalool and HETP calculated for the total reflux operation at 333 K and 8.8 MPa are shown in Figure 4. The selectivity increased with an increase in CO_2 flow rate. Higher selectivity up to 1700 was obtained. Therefore, the length of theoretical stage decreased as CO_2 flow rate increased. Calculated HETP decreased from 4 m to 0.2 m as the CO_2 flow rate increased for a model mixture of limonene and linalool.

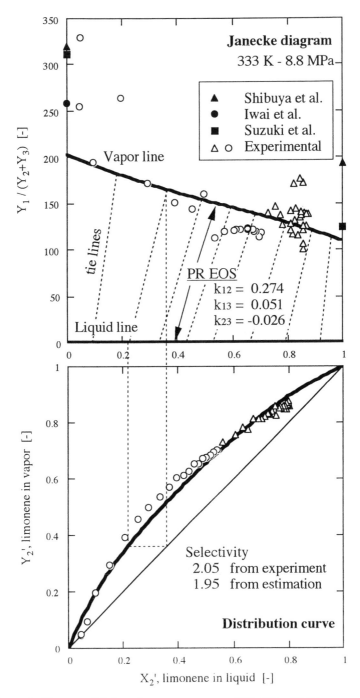

Figure 2   Phase equilibrium for ternary system of (1) CO_2 - (2) Limonene -
(3) Linalool at 333K and 8.8 MPa.

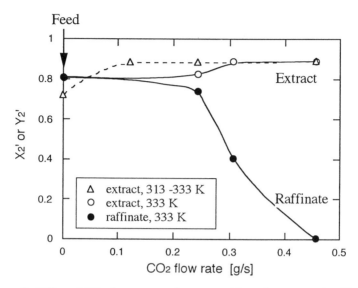

Figure 3   Effect of CO_2 flow rate on the composition of extract and raffinate for limonene for total reflux operation at 8.8 MPa.

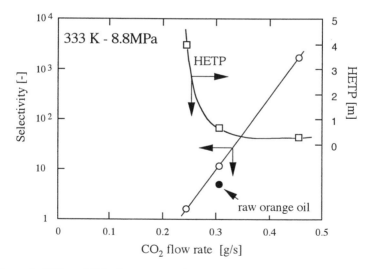

Figure 4   Effect of CO_2 flow rate on the selectivity and HETP for total reflux operation at 333 K and 8.8 MPa.

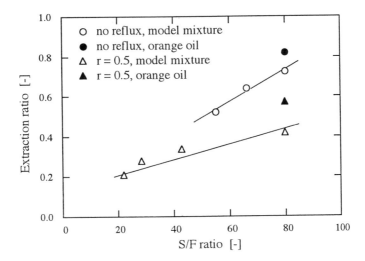

Figure 5   Effect of S/F ratio and reflux ratio, r, on the extraction ratio for continuous operation at 333 K and 8.8 MPa.

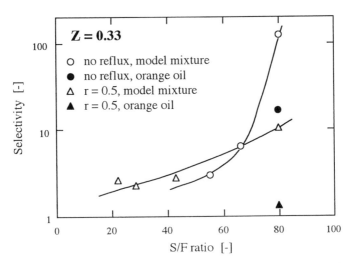

Figure 6   Effect of S/F ratio and reflux ratio on the selectivity for continuous operation at 333 K and 8.8 MPa.

When orange oil was used as feed, the selectivity was smaller than that of the model mixture. This may be due to differences of limonene contents in feed and some interaction among other solutes.

Continuous operation. The column was operated with countercurrent flow where CO_2 was a continuous phase and feed was a dispersed phase. In this work, only the external reflux was applied to the continuous operation at 333 K and 8.8 MPa. As discussed above, the flow rate of CO_2 solvent is an important factor to design the extraction process. During continuous operation, the solvent-to-feed (S/F ratio) was used as a flow rate factor, and the extraction was carried out at a constant feed flow rate of 0.42 g/min at various CO_2 flow rates.

Solvent-to-feed ratio and reflux. Effects of the reflux ratio on extraction ratio as a function of the S/F ratio are shown in Figure 5. The solid and open symbols are the values where raw orange oil and the model mixture, respectively, were used as feed. The extraction ratio is defined by the flow rate of limonene in the extract to that in the feed, indicating recovery ratio of limonene in the extract. The extraction ratio increased with the increase in the S/F ratio and was larger for the raw orange oil than the model mixture. The extraction ratio decreased by applying the external reflux.

Figure 6 shows the effect of the reflux ratio on the selectivity, b, as a function of the S/F ratio. The selectivity also increased with the increase in S/F ratio. The selectivity was larger for the operation with reflux at smaller S/F ratio. However, the external reflux reduced the selectivity at larger S/F ratio. As a result of the calculation of phase equilibrium for the system of CO_2 - limonene - linalool by PR-EOS as discussed above, the selectivity was about 2.0 at 333 K and 8.8 MPa. As the calculated selectivity corresponds to the simple extraction, those high selectivities shown in Figure 6 are a result of the mass transfer induced by countercurrent contact between the liquid phase feed and the vapor phase CO_2. Consequently, there is multistage for the separation in the column. On the other hand, the selectivity of the raw orange oil was lower than that of the mixture. These low selectivities may be due to low stage efficiency induced by high terpene contents and interaction among solutes.

Reflux ratio and raffinate flow rate. As shown in the above section, adding the external reflux reduced the selectivity at a higher S/F ratio. This result may be caused by the comparison at a constant S/F ratio. In fact, there is an intimate relation between the reflux ratio and the raffinate flow rate. Figure 7 shows the effect of the reflux ratio on the selectivity at various feed inlet positions as a function of the raffinate flow rate at constant S/F ratio of 80. As shown in this figure, the increase in reflux ratio leads to an increase in the raffinate flow rate at constant S/F ratio. Therefore, this indicates that the extraction ratio of limonene in the extract is reduced by an external reflux, resulting in lower selectivity. By applying the external reflux at a constant S/F ratio, the solute amounts staying in the rectification column may increase, then the actual solvent to feed ratio is decreased, where liquid phase is a mixture of feed and refluxed solute. If we operate at a constant raffinate flow rate, where the S/F ratio is not constant, the reflux contributes the desired separation.

Feed inlet position and reflux. The effect of reflux ratio on the selectivity characterized by feed inlet position is shown in Figure 8. Except for the inlet position of 0, the selectivity decreased with an increase in the reflux ratio at all feed inlet positions. The effect of the feed inlet position is not obvious. However, since these selectivities are mainly contributed from the raffinate composition, a larger decrease in limonene concentration in the raffinate shows larger selectivity. Feed inlet position of 0 indicates the length of enriching section is largest. This enriching effect is barely observed, where the selectivity increased from 1.8 to 2.1 as the reflux ratio

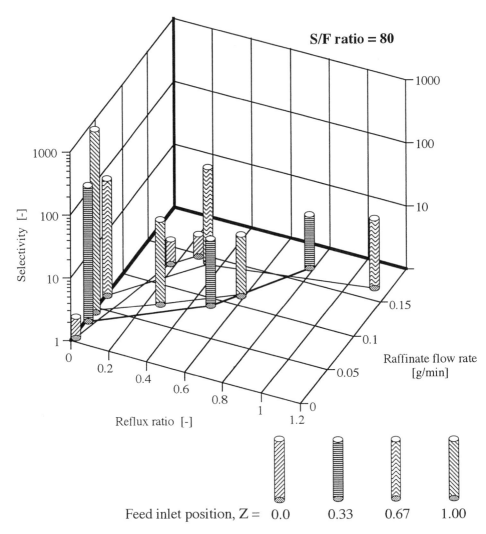

Figure 7   Relation between reflux ratio and raffinate flow rate for continuous operation at 333 K and 8.8 MPa.

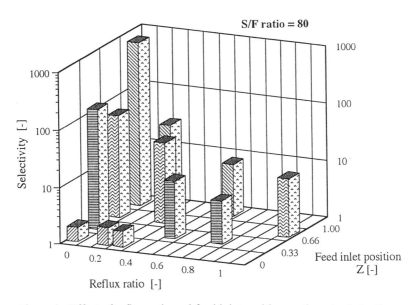

Figure 8   Effect of reflux ratio and feed inlet position on the selectivity for continuous operation at 333K and 8.8 MPa.

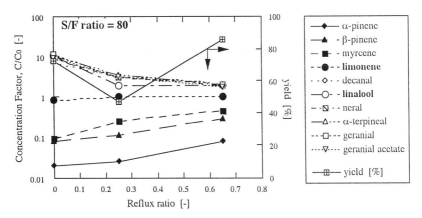

Figure 9   Concentration factor and yield of oxygenated compounds as a function of reflux ratio at 333 K and 8.8 MPa.

increased from 0 to 0.3. However, the effect of enrichment is absolutely smaller than that of stripping because of a higher limonene content in feed, resulting in lower selectivity. Consequently, the stripping section is more influential than the enriching section for the citrus oil processing.

Citrus oil processing. Cold-pressed orange oil from Brazil was processed at 333 K, 8.8MPa, S/F ratio of 80, and the feed inlet position of 0.33. Effect of reflux ratio on concentration factor of each major components in citrus oil and yield of oxygenated compounds in the raffinate is shown in Figure 9. The concentration factor, C/C_0, is defined by the concentration in raffinate divided by that in feed. The solid symbols are terpenes, which are desired to be smaller than unity, while the open symbols are oxygenated compounds, which are desired to be larger than unity. Oxygenated compounds were concentrated for no reflux operation at higher S/F ratio of 80. Although the yield, defined by the flow rate of oxygenated compounds in the raffinate to that in feed, was scattered from 45 to 85 %, the results are within the error of the operating material balance.

Because commercial requirement of the concentration factor is 5 - 10, these concentration factors of oxygenated compounds obtained at no reflux and S/F ratio of 80 are sufficient. To attain higher concentration factor for our bench scale apparatus, it may be necessary to operate at larger S/F ratio or to bubble CO_2 into the raffinate while keeping a certain height of interface between vapor and liquid. In this work the inlet position of CO_2 was set at the height over the interface to avoid this. On the other hand, a longer column may improve the separation at reasonable S/F ratio.

Conclusions

Supercritical fluid extraction of cold-pressed citrus oil and a model mixture of 80 % limonene and 20 % linalool was carried out at both total reflux operation and continuous countercurrent operation. The flow rate of CO_2 was most important to remove the terpenes from citrus oil. The increase of the CO_2 flow rate or S/F ratio gave a higher selectivity between terpenes and oxygenated compounds. Under a higher constant S/F ratio, the effects of reflux ratio and feed inlet position were investigated at 333 K and 8.8 MPa. The effect of the enriching section on the selectivity was smaller than that of the stripping section. Therefore, it may be preferable to provide a longer stripping section for citrus oil processing. Increasing the reflux ratio gave lower selectivity for an operation at constant S/F ratio.

Commercial desired product was obtained without decreasing the yield at S/F ratio of 80 and in no reflux operation. Oxygenated compounds were selectively fractionated in the raffinate and terpenes in feed was selectively removed in the extract. Extraction process operated at a higher S/F ratio will make a more valuable perfume product.

Acknowledgment

The authors are grateful to Mr. K.Maeda for experimental assistance. This work was supported by a Grant-in-aid for Scientific Research (No. 04238106) from the Ministry of Education, Science, Sports and Culture, Japan. We thank JSPS Research Fellowships for Young Scientists (No. 2362) to carry out this work. We also thank Givaudan-Roure Flav. Ltd. for supplying the citrus oil.

Literature Cited

1.   Stahl, E.; Gerard, D. *Perfumer and Flavorist* **1985**, 10, 29-37.
2.   Temelli, F.; Chen, C. S.; Braddock, R. *J. Food Tech.* **1988**, 42, 145-150.
3.   Yamauchi, Y.; Saito, M. *J. Chromatography*, **1990**, 505, 237-246.

4.   Barth, D.; Chouchi, D.; Porta, G. D.; Reverchon, E. and Perrut, M.*J. Supercrit. Fluids*, **1994**, 7, 177-183.
5.   Chouchi, D.; Barth, D.; Reverchon, E.; Della Porta, G., *Ind. Eng. Chem. Res.* **1995**, 34, 4508-4513.
6.   Perre, C.; Delestre, G.; Schrive, L.; Carles, M. *Proc. 3rd Int. Sym. Supercritical Fluids*, **1994**, 2, 465-470.
7.   Gerard, D. *Chem. Ing. Tech.* **1984**, 56, 794-795.
8.   Matos, H. A.; Azevedo, E. G. D.; Simoes, P. T.; Carrondo, M. T.; Ponte, M. N. D. *Fluid Phase Equilibria* **1989**, 52, 357-364.
9.   Giacomo, G. D.; Brandani, V.; Re, G. D.; Mucciante, V. *Fluid Phase Equillibria* **1989**, 52, 405-411.
10. Shibuya, Y.; Ohinata, H.; Yonei, Y.; Ono, T. *Proc. Int. Solv. Extr. Conf.* : York, **1993**, 684-691.
11. Suzuki, J.; Nagahama, K. *Kagaku Kogaku Ronbunshu*, **1996**, 22, 195-199.
12. Iwai, Y.; Hosotani, N.; Morotomi, T.; Koga, Y.; Arai, Y. *J. Chem. Eng. Data*, **1994**, 39, 900-902.
13. Peng, D.-Y.; Robinson, D. B. *Ind. Eng. Chem. Fundam.*, **1976**, 15, 59-64.
14. Sato, M.; Goto, M.; Hirose, T. *Ind. Eng. Chem. Res.*, **1996**, 35, 1909-1911.
15. Sato, M.; Goto, M.; Hirose, T. *Ind. Eng. Chem. Res.*, **1995**, 34, 3941-3946.

EXTRACTION AND CHROMATOGRAPHY

Chapter 11

The Future Impact of Supercritical Fluid Chromatography on Packed Columns, Modified Fluids, and Detectors

L. T. Taylor

Department of Chemistry, Virginia Polytechnic Institute and State University, Blacksburg, VA 24061–0212

Supercritical fluids possess many of the attributes necessary for high performance chromatography. By changing the density of the mobile phase with a change in temperature and/or pressure, one can significantly change the observed chromatographic characteristics in an SFC separation. Polar compounds are widespread and are often nonvolatile and thermally labile. These properties make analysis by GC impossible without derivatization, and make method development for LC complex. SFC does not require solutes to be volatile, since separations are carried out at low temperatures, and method development is straightforward.

The future of SFC will focus more on packed columns, modified CO_2, and a host of either universal, element specific, or spectrometric detectors. Consequently, SFC will become more viewed like HPLC and will, in fact, replace HPLC in a number of applications because supercritical fluids have both better mass transport properties than liquids and are less harmful to the environment than liquids. The relatively poor solvating power of CO_2 has dictated the use of primary and secondary modifiers in the mobile phase. These additives as well as packed columns which have much higher decompressed flow rates place new demands on the employment of detectors in SFC.

A supercritical fluid exhibits physico-chemical properties intermediate between those of liquids and gases. Mass transfer is rapid with supercritical fluids. The diffusion coefficient is (in the vicinity of critical point) more than ten times that of a liquid. Density, viscosity, and diffusivity are dependent on temperature and pressure. The viscosity and diffusivity of the supercritical fluid approach that of a liquid as pressure is increased at fixed temperature. Diffusivity will increase with an increase in temperature at fixed pressure; whereas, viscosity decreases (unlike gases) with a temperature increase. Changes in viscosity and diffusivity are more pronounced in the region of the critical point. Even at high pressures (300-400 atm), diffusivity is 1-2 orders of magnitude greater than liquids. Therefore, the properties

134

of gas-like diffusivity, gas-like viscosity, and liquid-like density combined with pressure-dependent solvating power have provided the impetus for applying supercritical fluid technology to analytical separation problems.   Supercritical fluid chromatography (SFC) is an analysis technique that uses supercritical fluids as the mobile phase.   Liquid chromatography (LC-like separations that exhibit more gas chromatography (GC)-like figures of merit such as high speed, high resolution, and multiple detection options are characteristic of SFC with packed columns.   Open tubular column SFC is an extension of GC to larger, low volatile, and more thermally stable molecules.   The approach to methods development varies greatly depending on the column type.   This review will describe in a cursory manner both philosophies.

History

The use of a supercritical fluid as a chromatographic mobile phase was first reported in 1962 by Klesper, Corwin and Turner.[1]   Chlorofluorocarbon mobile phases were used to separate nickel porphyrins.   Four years later the first SFC chromatogram appeared, employing isobaric (i.e. constant pressure) conditions.[2]   It was not until 1970 that the use of pressure programming in an SFC experiment was demonstrated.[3]   In 1981-82, Hewlett-Packard introduced SFC instrumentation (e.g. modified liquid chromatograph) for packed column SFC at the Pittsburgh Conference.   Concurrent with this event was the first report on the use of open tubular, capillary columns in SFC.[4]   In 1986, several vendors at the Pittsburgh Conference introduced capillary SFC instrumentation.   A book[5] entitled Analytical Supercritical Fluid Chromatography and Extraction was published in 1990 which was a great aid to workers in the field.   More recently Berger has published a book solely on packed column SFC.[6]

Berger has stated that SFC may be unusable with 30% of all molecules and that laboratories will probably eventually have 20% as many SFC instruments as LC instruments.   SFC is considered to possess inferior figures of merit compared to GC but SFC is more widely applicable.   On the other hand, SFC possesses superior figures of merit compared to LC but SFC is less applicable than LC.   It should be noted that all the control parameters available in both GC and LC are available and useful in SFC (e.g. mobile phase composition and identity, temperature, pressure, flow, and stationary phase identity).   Methods development should be more straightforward in SFC than HPLC.

Figure 1 compares the HPLC and SFC traces of biphenyl and pyrene at the optimum average linear velocity for each.[7]   An octadecylsilica (ODS) reversed phase column, 10 cm x 4.6 mm and 5 μm particle diameter, was used.   The SFC separation was completed in less than two minutes; whereas, the HPLC separation required over four minutes.   Experimentally derived van Deemter plots from HPLC and SFC elution of pyrene are shown in Figure 2.   For the same packing material, the minimum height equivalent to a theoretical plate (HETP) was the same regardless of whether a liquid mobile phase or supercritical fluid mobile phase was employed.   The optimum SF linear velocity, on the other hand, was more than double the HPLC optimum linear velocity.   Such time savings relative to GC, on the contrary, do not exist with SFC since gases afford much higher optimum linear velocities than SFs.   From a speed and efficiency standpoint, GC should be the first choice, SFC the second choice, and when neither of these techniques are applicable, HPLC should be selected.

In terms of economic and environmental issues (e.g. solvent price and disposal), SFC may also be preferred over HPLC.   This advantage in analysis is

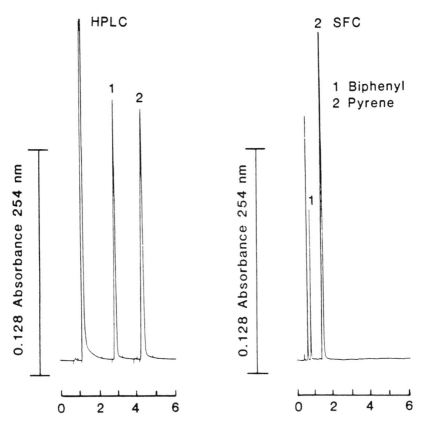

Figure 1.   Chromatograms of the separation of biphenyl (peak 1) and pyrene (peak 2) at the optimum average linear velocity for HPLC and SFC.   An ODS reversed-phase column (10 cm by 4.6 mm) and a 5-μm particle diameter were used.   For HPLC the solvent was acetonitrile and water (70:30) at 1.0 cm^3/min, linear velocity 0.13 cm/sec, and column pressure drop 62 bars. For SFC, the carbon dioxide flow rate was 2.5 cm^3/min., linear velocity 0.40 cm/sec, column pressure drop 14 bars, and average column pressure 165 bars.

particularly striking when solvent usage and sample throughput for SFC and HPLC of felodipine are compared, (Table I and II)[8]   When the SFC/UV system was used for analysis, sample throughput was increased by 60% over an analogous HPLC separation. Although more total mobile phase was used for SFC, only 6% of the SFC mobile phase was actually disposable solvent waste. The remaining 94% was carbon dioxide gas which was vented to a hood. The disposal cost of 100% organic solvent (non-chlorinated) vs. water/organic solvent mixtures also illustrates another advantage of the SFC assay.   The most common procedure for solvent waste disposal is combustion in large scale manufacturing furnaces.   Such furnaces typically combust 45,000 gallons of solvent waste/h. Since water/organic mixtures, generated from HPLC analysis, produce less heat (<3000 B.T.u./lb.) upon combustion, the resulting cost of disposal to the waste source is higher. Conversely, 100% organic solvent disposal (generated by SFC) has a higher fuel value (9500 B.t.u./lb.) and, therefore, its cost of disposal is less.

Theoretical Considerations

While many workers have extolled the virtues of both open tubular and packed columns for SFC, experimental comparisons are generally lacking, wherein typical operating conditions and identical parameters (where feasible) are employed. Schwartz et al.[9], however, have made an excellent comparison of these two types of columns based on theoretical considerations. Figure 3 shows theoretically generated van Deemter curves from this study for both packed and open tubular columns. At linear velocities higher than optimum, the efficiency of the column decreases as evidenced by the higher HETP.   The effect of higher linear velocity on column efficiency was predicted to be more pronounced for larger molecules due to their higher elution density than for smaller molecules. For the same typical molecules, an open tubular column of 50 μm internal diameter was suggested to show an even more dramatic increase in HETP at increased linear velocities; while, a packed column was predicted to be more tolerant of increases in linear velocity of the mobile phase. This implies that higher working linear velocities can be used with a packed column without significant loss of efficiency, which would ultimately lead to faster analysis times provided sufficient resolution is available. In practice o.t. columns are always operated at the outset at linear velocities greater ($5u_{opt}$) than the optimum in order to achieve reasonable retention times.   In comparison reasonable packed column retention times are obtainable at slightly less than the optimum ($0.2u_{opt}$).[10]

The effect of increased linear velocity on column efficiency can be very significant when passive fixed diameter restrictors located at the column outlet are used. Under these conditions higher operating density is achieved only at the expense of a greater pumping rate. The pump delivers whatever flow is required to achieve the column head pressure setpoint. In other words, density and flow rate are coupled when fixed restriction is employed. This situation has been described as operation in the upstream mode. Upstream control is most often used in situations that require very low flow rates. A worst case scenario, in the upstream mode developes during an SFC run with density programming. The optimum linear velocity is decreasing as the density increases because conditions are becoming more liquid-like; whereas, the operating linear velocity is increasing as density increases because the pump is working harder. Under pressure programming conditions, if one starts out at the optimum mobile phase velocity, one could easily be at several times the optimum velocity by the end of the run. For example, with a 7 m x 50 um i.d. column at 40°C, the linear velocity goes from 1.3 cm/sec at 0.47 g/mL to 10.2 cm/sec at 0.96 g/mL. If the column were operated at 100°C, the increase in linear velocity would be 2.3X.[5]

Table I

Packed Column SFC vs. HPLC[+]

	SFC	HPLC
Total Retention Time, t_R	4.34 (0.3%)	9.23 (0.5%)
Hold-up time, t_m (min.)	1.26	1.52
Retention Factor, k'	2.44	5.07
Peak width at Half-Height, w_h (min)	0.089	0.25
Plate Number, N	13115	8519
Plate Height, H (mm/plate)	0.017	0.029
RSD (Peak Area)	1.1%	1.2%

SFC analysis conditions: 25 cm x 4.6 mm i.d. Hypersil Si column; temperature, 45°C pressure, 300 bar; flow rate, 2 mL/min.; injection volume, 5 µL. Felodipine concentration was 1 mg/mL.

[+] All peak parameters were calculated based upon 5 replicate injections of a felodipine standard. RSD values are given in parentheses.

Table II

Solvent Usage Comparison for Analysis of Felodipine
Packed Column SFC vs. HPLC

	Packed Column SFC/UV	Analytical Scale HPLC/UV
Mobile Phase	6% (v/v) methanol-modified CO_2	Acetonitrile-methanol/50 mM phosphate buffer (pH 3) (40:20:40, v/v/v)
Samples Analyzed/h	10 (6 min. run time)	4 (15 min. run time)
mL Mobile Phase Used/h	120	90
mL Disposable Waste/h	7.2	90
Mobile Phase Disposal Cost/55 gal.	$48+	$175+

[+]Disposal cost obtained from Solid Waste Management, Merck Research Laboratories, West Point, PA

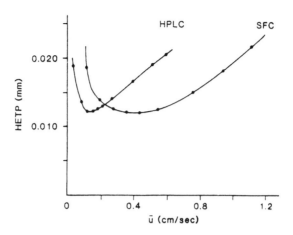

Figure 2. Van Deemter plots for chromatographic data for HPLC and SFC elution of pyrene; HETP is height equivalent to a theoretical plate. Conditions: 10 cm x 4.6 mm i.d. packed column, C_{18} bonded silica particles (Hypersil) 5 μm. HPLC, K=2.85, 30% H_2O in CH_3CN, 40°C. SFC, CO_2 at 0.8 g/mL, K=2.30. From L. G. Randal, ACS Symp. Ser., 250, 135 (1984).

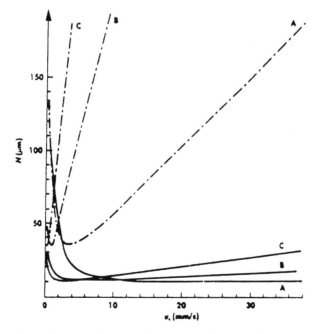

Figure 3. Theoretical van Deemter plots for packed and capillary columns. (Reproduced with permission from reference 9. Copyright 1987 Advanstar Communications.)

High pressure electronically controlled micrometering valves are becoming quite popular in packed column SFC. These backpressure regulators which are termed variable restrictors allow flow rates to be adjusted to constant levels at different densities (pressures). Because the pressure is controlled after the column, the mode of operation is termed the downstream mode. Capillaries and micropacked columns require low flow rates and have such low dead volumes that it is presently impossible to perform downstream control with existing equipment. The reader is referred to reference 5 for the designs of several variable restrictors. Currently, pre-restrictor detectors (e.g. UV, FTIR) and analytical scale packed columns are applicable to variable restriction. Much of the future acceptance of SFC by separation scientists rests upon the universal incorporation by workers in the field of variable restrictors where mass flow and pressure can be decoupled. Without a variable restrictor, it becomes extremely difficult to deconvolute the individual effects of changing the t_0, fluid linear velocity, retention factor and column efficiency.

Fluids and Modifiers

In lieu of increasing the solvent power of the SF by increasing the density, the characteristics of SF phases can be varied by the addition of polar compounds (i.e. modifier) to the primary fluid. Retention factors are not only a function of modifier properties but also of modifier concentration. The use of binary phase systems offers great flexibility since the modifier identity and concentration can be easily changed. Both isocratic and gradient delivery have been employed. Methanol is by far the most commonly used modifier in SFC. Acetonitrile, which has a higher polarity index than that of methanol, is less soluble in CO_2 than methanol and therefore is less frequently used. Water is even less soluble in CO_2 than acetonitrile; therefore, when water is used as a modifier in CO_2, the mobile phase is usually saturated with water.

The disadvantage of adding even small amounts of modifier to carbon dioxide is that, in most cases, flame ionization, for example, can no longer be used as a detector. This presents a problem for compounds that do not have UV chromophores. Another disadvantage of using a modified mobile phase for SFC is that the critical parameters (T_c, P_c) of the binary phase are higher than the critical parameters of the pure fluid. Critical parameters can be calculated using equations of state, however, it has been reported that calculated parameters such as these have been found to differ by 20% from empirical measurements.[11] If the compounds of interest are thermally labile, working at the higher temperatures required in order to produce a supercritical phase may be a disadvantage. In order to achieve a separation, supercritical conditions are not absolutely required if operating at relatively high CO_2 densities and using an ultraviolet detector. Subcritical or near-critical chromatography has been demonstrated to be effective on numerous occasions with primarily packed columns.[12]

The effects of modifiers in SFC are: (1) coverage of active (silanol) sites, (2) swelling or modification of the stationary phase, (3) increasing the density of the mobile phase, and (4) increasing the solvent strength of the mobile phase.[13] The mechanism of action of modifiers, however, is somewhat ambiguous and competing mechanisms (i.e. stationary phase effects vs. mobile phase effects) have been proposed. It has been reported that, in contrast to packed columns, o.t. columns do not show the drastic changes in retention factors or peak shape upon addition of small (<2%) amounts of modifier.[5] These less drastic differences were attributed to the differences in the degree of deactivation of the packed column stationary phase as compared to the o.t. column stationary phase. An o.t. column has a smaller number of active sites present; therefore, less active sites are present for the modifier to deactivate. Open tubular columns do have active sites, as it has been reported that column coating efficiencies for small diameter open tubular columns range from 20-80% depending on the stationary phase.[13]   However, Berger et al. contend that modifiers produce about the same results in o.t. and packed columns. They reported that the retention of polyaromatic hydrocarbons (PAHs) decreased 15-32% on a variety of packed columns and 26-28% on o.t. columns when approximately 2% of 2-propanol or methanol was added to the mobile phase.[13]

The modifier may be further altered by introducing low concentrations of a very polar compound. When a secondary modifier is added to the mobile phase via the primary modifier it is sometimes referred to as an additive.[14]   Acetic, citric, chloroacetic, dichloroacetic, trichloroacetic, and trifluoroacetic acid have been studied as acidic additives, while tetrabutylammonium hydroxide (TBAOH) and isopropylamine[13] have been studied as basic additives.

Berger et al.[15] have separated mono-, di-, and trihydroxybenzoic acids on cyanopropyl, diol, and sulfonic acid columns. When pure carbon dioxide was used none of the acids eluted from the columns. When methanol was added to the mobile phase, some of the acids eluted, but with very bad peak shapes. The addition of citric acid to the mobile phase allowed for the separation of ten mono-, di-, and trihydroxybenzoic acids in approximately 1.5 minutes with much improved peak shape. The authors concluded that the most predominant action of additives is to improve the solubility of the solute in the mobile phase and to suppress the ionization of very polar solutes. They also concluded that very polar additives interact with the active sites on a column so strongly that they can not be removed by the solutes, therefore serving to further deactivate the column.

For SFC systems the gas supply is simply a laboratory sized cylinder from any number of manufacturers. Carbon dioxide, by virtue of its moderate critical parameters, high purity (SFC grade), and low cost is the most common SF being used today - but many other potentially useful fluids are available. From the cylinder which contains mostly liquid with a gas headspace, the latter is drawn from the tank to the pump where it is pressurized to the desired value. The head pressure of the gas forces liquid up through the dip tube which extends to the bottom of the cylinder from the valve. The liquid then travels through the valve, plumbing, and into the chromatographic pump. It has been common practice to add a helium headpressure to the supply cylinder so that the cylinder pressure is greater than the vapor pressure of the fluid. The helium is intended to aid in pushing the liquid at the bottom of the tank up and out. Helium, however, dissolves in CO_2 which gradually changes its composition. Future SFC experiments will probably not use helium padded fluids since with properly designed pumping systems padding is not necessary.   Two major types of pumps are found in SFC instruments: syringe and piston. Syringe pumps have fixed volumes, therefore, dual syringe pump arrangements are employed whereas one is being emptied,the other one is being filled. Piston pumps are only limited by the liquid volume of the gas-liquid supply cylinder. However, it is necessary to cool the piston heads in some manner so that only the non-compressible liquid phase is pumped. The pump controller on most SFC instruments allows several gradients such as CO_2 pressure, CO_2 density, temperature, and mobile phase composition to be programmed into the method.

Modifiers can be introduced into chromatographic systems in primarily two ways. First, premixed tanks of modified fluid can be purchased with a variety of fluids and modifier percentages, and are usually sold as weight % modifier in pure

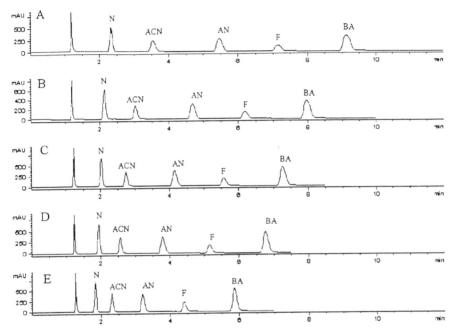

Figure 4.   Separation of PAH's using five different types of CO_2.   Chromatogram A: Level 1 HHS-CO_2, Chromatogram B: Level 2 HHS-CO_2, Chromatogram C: Level 3 HHS-CO_2, Chromatogram D: Level 4 HHS-CO_2, Chromatogram E: Level 5 Pure CO_2.   Peak identity as follows: (I) Impurity (N) naphthalene, (ACN) acenaphthylene, (AN) anthracene, (F) fluoranthrene, and (BA) benz(a)anthracene. Description of CO_2 samples follows:

Description	Cylinder	Cylinder + Contents Weight (kg)
Pure CO_2	100% CO_2	N/A
Level 1	Full Tank of Helium Headspace CO_2	40.71
Level 2	3/4 Filled Tank Helium Headspace CO_2	36.24
Level 3	1/2 Filled Tank Helium Headspace CO_2	31.78
Level 4	1/4 Filled Tank Helium Headspace CO_2	27.51

(Reproduced from reference 17. Copyright 1996 American Chemical Society.)

fluid. Poor mixing and fluctuating modifier content in premixed gas cylinders have long been suspected as a source of erratic chromatographic behavior in SFC. A recent model by Schweigardt and Mathias[16], incorporating a modified Peng-Robinson equation of state (EOS) with three binary interaction coefficients, was used to model vapor-liquid equilibrium (VLE) changes as the liquid volume in a premixed cylinder is depleted. They specifically modeled the dimensions of the 1.04 ft^3 aluminum tank from which most SFC mobile phases are provided. The model predicted that methanol concentrations in the mobile phase could increase two-fold during the usage-life of the cylinder. As the liquid volume in the cylinder is depleted, the total gas volume above the liquid increases. Since the primary fluid (carbon dioxide) has a much higher vapor pressure than the modifier (methanol) it preferentially moves into the vapor phase, occupying the newly created volume. The amount of methanol leaving the liquid phase was considered negligible compared to CO_2. Thus, the resulting shift in mass balance leaves the methanol at higher percent compositions in the liquid as the cylinder continues to be depleted. Although premixed cylinders allow some exploratory work with a single pump, the results cannot be reproduced. Pre-mixed binary fluids should therefore be avoided in SFC.

The same VLE shifts that result in increases in modifier concentration in premixed cylinders may result in higher concentrations of impurities in pure CO_2 cylinders as their liquid volumes are reduced during use. Additionally, helium is known to be quite soluble (0.04 mole fraction at 2000 psi) in liquid CO_2 under the pressure-temperature conditions typically found in He headspace pressurized cylinders. It has been observed that as pure CO_2 tanks with He headspace are depleted, there are retention time shifts as a result of He as opposed to CO_2 preferentially moving into the gas volume above the liquid level.[17] The retention times of PAH's increased when helium headspace carbon dioxide was used as a carrier fluid relative to pure carbon dioxide because of the reduction in CO_2 density (Figure 4). The increased retention times were affected, as predicted, by the level of liquid carbon dioxide present in the helium headspace carbon dioxide cylinder. As more liquid carbon dioxide was removed from the cylinder, the effect of helium on the solvating power of CO_2 was reduced because the relative amount of helium dissolved in the liquid phase decreased. Furthermore, the effect of helium headspace carbon dioxide was investigated with methanol-modified carbon dioxide mobile phases for the analysis of steroids. Surprisingly, the relative solubility of helium in carbon dioxide again resulted in longer retention times when compared to pure carbon dioxide as the liquid level of carbon dioxide decreased.

A second way to add modifier to a chromatographic system is to use a two pump system where one pump delivers the pure fluid and the other pump delivers the liquid modifier. The two fluid streams are mixed in a volume-volume ratio to form the mobile phase. This method of preparing the mobile phase provides the greatest flexibility in that the mobile phase is mixed in-line. It is also an accurate way to add modifier, since the flow rate of each pump is known. Usually a reciprocating piston pump is used to deliver the pure fluid, while a syringe pump is used to deliver the modifier. One of the main concerns in mixing a liquid modifier and the supercritical fluid is that the compressibilites of the two fluids are different. The compressibilities of the fluids are of a special concern if a pressure gradient is to be used. For example, if a pressure gradient at constant composition is required, then as the pressure increases, the relative speed of the syringe pump delivering CO_2 decreases compared to the speed of the syringe pump delivering a less compressible organic modifier.[18] Since the modifier concentration in SFC is generally small, the main fluid pump and the modifier pump also operate at different frequencies.

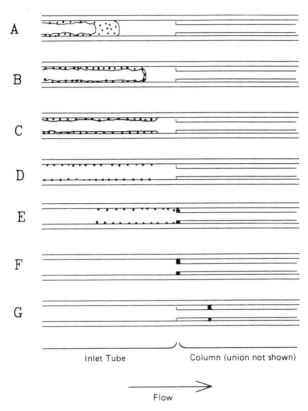

Figure 5.   Representation of sample spreading due to inlet flooding, and the refocusing effect possible with the use of an uncoated inlet tube.   (A) Solute is carried over a length of inlet tubing (or column) by the liquid injection solvent. (B,C) Sample spreading continues until injection solvent is depleted (or sufficiently diluted), (D) Solvent-free solute is left spread over what was the flooded zone. (E,F) If flooding occurs on an uncoated inlet tube, solute migration will begin at mobile-phase strength too low for significant migration on the column stationary phase. Solute reaching the stationary phase is refocused into a narrow band. (G) Solute migration along the column begins as the mobile-phase strength is raised further.
(Reproduced with permission from *Analytical Instrumentation Handbook;* Chester, T. L.; Marcel Dekker: New York, 1990; Figure 7. Copyright 1990 Marcel Dekker.)

The solubility of the modifier in the supercritical fluid is also important. If the solubility of the modifier is exceeded, a two-phase system will result, and the effect of a two-phase system on chromatography may be detrimental. A UV detector often gives an indication of whether a two-phase system is present or not, in that a very noisy signal results when two phases are present. High solubility is not required for SFC, but obviously chromatography becomes totally impossible without some finite solubility. Mixtures of two miscible fluids usually do not become instantaneously homogeneous after the mixing point. A packed bed of stainless steel balls downstream of the mixing point usually suffices as a mixing column.

Sample Injection

The most common injectors for SFC are high pressure valve injectors similar to those used in HPLC.[10] With these valves the sample is loaded at ambient pressure in a sample loop of defined size (100-500 nL). Direct full loop injections are the normal means of sample introduction in SFC with packed columns. After the sample is loaded into the sample loop, the valve is manually or pneumatically actuated. This operation places the sample loop in-line with the high pressure mobile phase flow. The loop is purged with liquid CO_2 which becomes supercritical upon entering the heated zone and column. Injections should be made at low pressure and ambient temperature. Complex phase behavior phenomena may exist during the injection process at the injector and at the head of the column since the sample injection solvent is almost always an organic solvent. Peak splitting/peak shoulders can be avoided by complete mixing of the injection solvent and fluid.

The most common method currently used for injection with open tubular columns is split injection. Several split injection methods have been employed. Dynamic flow split is the simplest and most popular, yet there are a number of problems associated with flow split injection. For example, viscosity variations among different samples can change the splitting characteristics making the use of external standards impractical. Accuracy and precision are usually much better with time-splitting injection than dynamic flow split injection, especially with internal standards. Chester and Innis[19] have reported that the key to successful quantitative analysis is the complete avoidance of splitting. If everything is done correctly, precision should be limited only by the filling of the injection valve loop and should match the precision of HPLC. They have developed an injection approach relying on the formation of a liquid film of the injection solvent on a retention gap, and the subsequent refocusing of the solutes from the flooded zone onto the head of the column. The entire operation is illustrated in Figure 5. With this approach volumes up to 0.5 µL were injected by Chester and Innis with little or no sacrifice of chromatographic performance while achieving relative standard deviations of less than 0.3% for the injected volume. Relative standard deviations for solute area were only slightly worse, depending on the peak shape, but are generally better than 2% for well-shaped, baseline-resolved peaks.

Packed columns pose much less of a sample injection problem owing to their greater loadability, and direct injection methods have, in fact, proven highly effective and reproducible for packed column use. In fact, for packed columns of 4.6 mm and 1.0 mm inner diameter, direct injection is as routine and reliable as found in HPLC injection. The only difference might be that the injection solvent can never be the mobile phase unless one is performing supercritical fluid extraction directly coupled to SFC (e.g. SFE/SFC). Large sample volume injections are, however, to be

Table III

Structures of Polysiloxane
Stationary Phases

Methyl-100

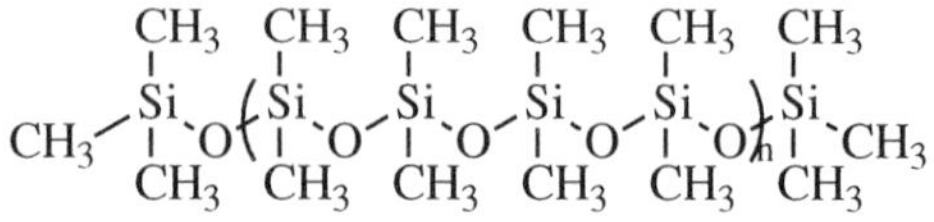

Octyl-50

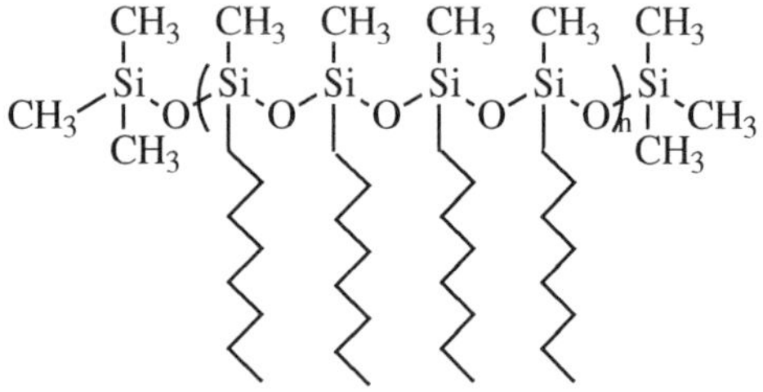

Phenyl-5
Phenyl-50

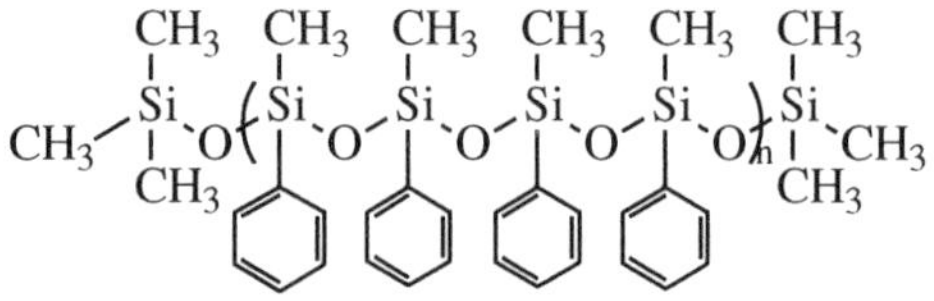

Biphenyl-30
Biphenyl-50

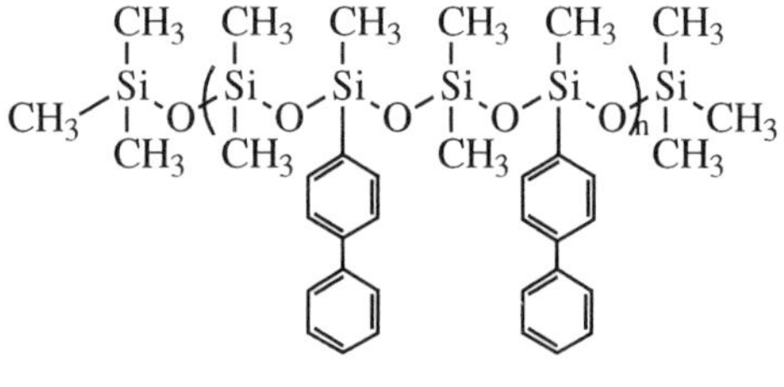

Cyanopropyl-25
Cyanopropyl-50

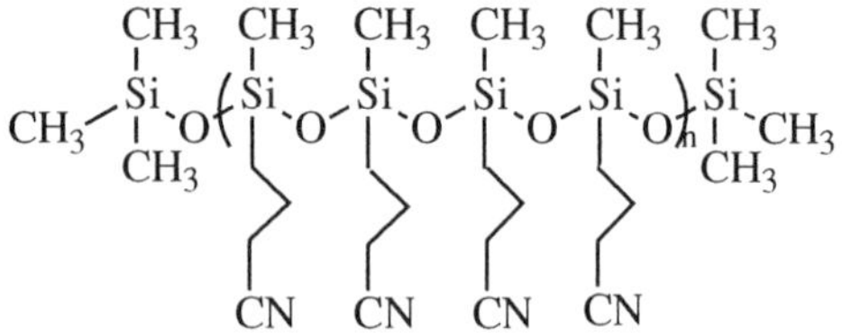

Liquid crystal

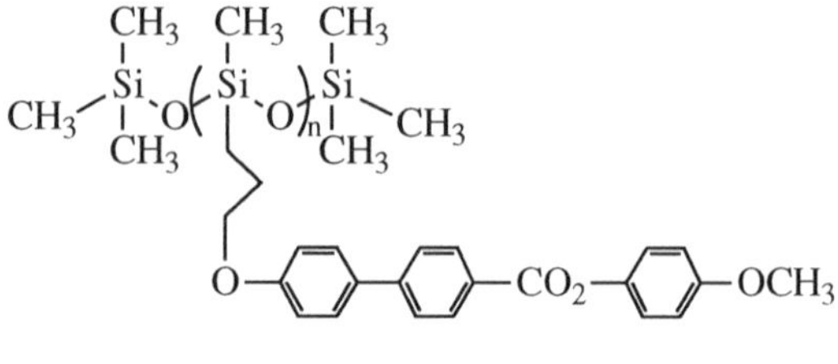

avoided unless a solvent elimination protocol is introduced prior to the separation. Excessive solvent may cause band broadening and peak splitting, and the properties of the stationary phase may change.

Stationary Phases

Open-tubular coated capillaries (50-100-μm i.d.), packed capillaries (100-500-μm i.d.), and packed columns (1-4.6-mm i.d.) have been used for SFC. Regardless of the column, certain requirements must be met: high efficiency, good selectivity, deactivated surfaces, and thermal stability. In addition, uniform and immobilized stationary phase films on the inner walls are required for open-tubular columns. Conventional GC open-tubular columns are not used in SFC because their column inner diameters are too large and their stationary phases, which have not been extensively cross-linked, migrate on the column under supercritical conditions. On the other hand, HPLC columns are routinely and interchangeably used for SFC.

Many silicone stationary phases developed for either GC or HPLC have been adopted for use in SFC. These include phases exhibiting all types of solute-stationary phase interactions and selectivities such as adsorption, dispersion, dipole-induced dipole, dipole-dipole, and size and shape. Table III lists some of the stationary phases which are available with o.t. columns. The most widely used polar phase in o.t. columns is cyanopropyl-polysiloxane.[20] Special note of the liquid crystalline stationary phase should be taken in Table III.[21] Separation is a function of the analyte's conformation or spatial arrangement and is temperature dependent. Figure 6 shows the type of separation of carbazole substitutional isomers that can be achieved on the liquid crystal phase versus a more conventional methyl phase. A dramatic enhancement in resolution over GC was demonstrated for selected geometrical isomers. The SFC elution was performed at 120°C where the liquid crystal phase was more ordered than at the 230°C elution temperature in GC.

In packed columns, the stationary phase is normally near monomolecular thickness and is polymerized and chemically bonded to the support. Particle sizes vary from 3 to 10 μm in diameter with pore sizes ranging from 100 to 300 Å which corresponds to a surface area of 100-300 m^2/g. Packed column stationary phases are silica based and may either be of the LC type or polymer coated type. The surface activity in the LC case is a serious limitation when mobile phases of low polarity (such as CO_2) are used. The silica surface activity appears to be directly related to (a) the number of silanol sites remaining after bonded phase application, (b) the accessibility of these silanol sites, and (c) the degree to which these residual silanols are covered with physically adsorbed mobile phase. With supercritical CO_2, silanol sites are essentially uncovered. Although polar organic modifiers such as methanol can be used to alleviate the active silanol problem, the interfering detector response of most modifiers often limits GC-type detector capabilities. Many workers have shown that conventional endcapping is not effective in that all silanol sites are not reacted, and at temperatures above approximately 120°C the endcapping reagent is not stable.

A more satisfactory solution to this dilemma, specially developed for SFC, involves the use of hydrosiloxane polymers that are coated and chemically bonded to a porous silica particle.[22] A surface coating technique similar to one previously developed for capillary GC was employed. The separation of a mixture (4 μg/μL each) of n-pentadecane, phenyl acetate, acetophenone, 2,6-dimethylaniline, and phenol on both a regular cyanopropyl and a cross-linked cyanopropyl coated packed column is shown in Figure 7. Peak shapes were much improved for the cross-linked phase, especially for 2,6-dimethylaniline, the most basic component. All polar

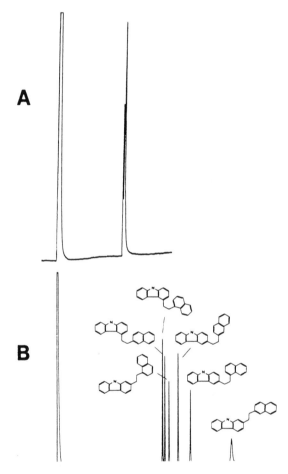

Figure 6.   Supercritical fluid chromatograms of six naphthylcarbazolylethanes on (A) SB-methyl-100 and (B) liquid crystal columns.   SFC conditions: density programmed from 0.25 g/mL to 0.658 g/mL at 0.0075 g/mL/min after a 10-min isoconfertic period; temperature held at 140°C.
(Reproduced with permission from reference 21. Copyright 1988 Preston Publications.)

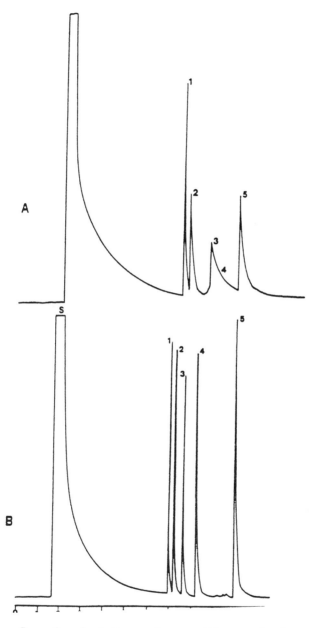

Figure 7.   Separation of polarity test mixture on (A) conventional cyanopropyl and (B) polymer coated cyanopropylcolumns with 100% CO_2 and flame ionization detection at 60°C. Peaks: (1) *n*-pentadecane, (2) phenyl acetate, (3) acetophenone, (4) 2,60dimethylaniline, (5) phenol.

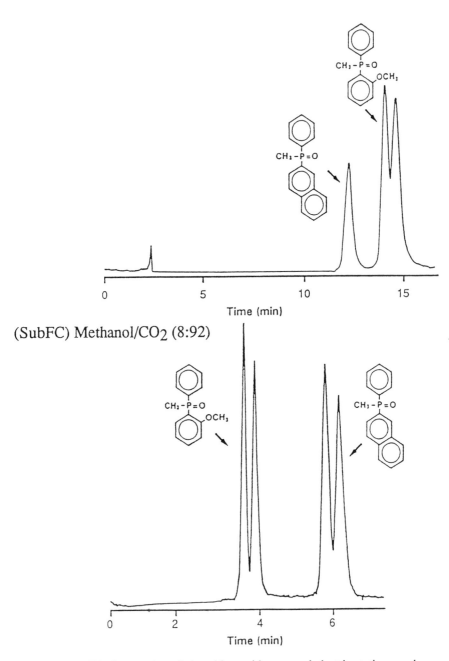

(SubFC) Methanol/CO$_2$ (8:92)

Figure 8.   Chiral separation of phosphine oxides on cyclodextrin stationary phase
employing liquid chromatography and subcritical fluid chromatography.
(Reproduced with permission from reference 23. Copyright 1989 Preston Publications.)

solutes exhibited some tailing on the regular cyanopropyl column. The difference in t_0 between the two separations can be attributed to the fact that the silica base, pore size, and pore size distribution are somewhat different for the two columns.

Packed column stationary phases are finding real significance in chiral separations which are important to the pharmaceutical industry.[23] The two most popular chiral stationary phases are the Pirckle phases where π-interaction and hydrogen bonding prevail and the cyclodextrin phases where an inclusion complex interaction is thought to dominate. Cellulose derivatives coated onto silica gel are also used. With packed columns most of the separations are performed under subcritical conditions since enantiomeric selectivity is usually enhanced at low temperature. Figure 8 shows the chiral separation of two phosphine oxides (e.g. 2-naphthyl and o-anilyl) employing a liquid mobile phase and a subcritical mobile phase[24] with a cyclodextrin stationary phase.

By combining the advantageous features of both packed and open tubular columns, packed capillaries open new possibilities in high performance SFC separations. Because of the inherent low flow rates, packed capillary columns have significant advantages over conventional packed columns for use in hyphenated systems like SFC-MS. Packed capillary column SFC will also provide remarkable economy in SFC operation, requiring a few orders of magnitude lower mobile and stationary phase consumptions as compared withconventional packed column SFC. Many excellent papers on the potential of packed capillary SFC have been published in recent years. W. Li et al.,[25] Y. Shen et al.,[26] E. Francis et al.,[27] and E. Ibanez et al.[28]

Detection

A major advantage of SFC over HPLC is that SFC is truly "detector friendly". Both HPLC and GC detectors have been successfully interfaced to SFC. Detection may occur before the restrictor using a closed cell design in which the fluid is maintained under pressure. Alternately, detection may occur after decompressin to a gas (i.e., post restriction).

The most useful SFC detector is flame ionization[29] when CO_2 is used as the mobile phase. The flame-ionization detector (FID) responds to most organic molecules with a detectin limit for carbon of 1-10 pg/s and a linear dynamic range of 10^6. Gas flow rate and restrictor placement are critical. The flame structure is larger for carbon dioxide than helium. Therefore, the collector position and voltage must be optimized. Because Joule-Thompson cooling accompanies CO_2 decompressin (e.g., expansion), the temperature of the FID should also be higher than used in GC. In addition to its sensitivity, FID is amenable to pressure programming, is easy to operate, does not respond to CO_2, and is applicable to both open-tubular and packed columns.

Electron-capture detection (ECD) is highly sensitive to compounds with high electron affinities. In SFC, the decompressed mobile phase flow rate must be low; consequently, open-tubular columns were demonstrated first. A make-up gas of 10% CH_4 and 90% Ar is necessary in order to enhance the diffusion of thermal electrons in CO_2 and to achieve a stable baseline. Density programming is possible for this concentration-sensitive detector. Both associative and dissociative electron-capture mechanisms are in operation; therefore, the detector temperature is critical.

Additional post-restrictor detectors have been shown to be feasible with SFC. Sulfur chemiluminescent detection is readily amenable to modified fluids and to all types of columns. Flame photometric detection is also sulfur specific, but response varies with CO_2 density and depends upon the sulfur content in the compound in contrast to sulfur chemiluminescent detection. Photoionization, ion mobility, thermionic, and evaporative light scattering have also been shown to be feasible for SFC.

The second most popular detector for SFC is ultraviolet detection. With packed columns this pre-restriction detection is simple; whereas, with o.t. columns the interface dead volume is a serious issue. Regardless of the column, certain trade-offs have to be made in flow cell design to achieve small flow cell volumes and relatively long flow cell pathlengths. These compromises are not severe for packed columns but for o.t. columns several strategies have been attempted such as on-column detection and pseudo on-column detection.[30] Neither of these strategies have proven very satisfactory and the failure of o.t. columns with UV is an area where improvements are needed. With pre-restrictor detectors, analysis is usually performed in the liquid state rather than the supercritical state because solute concentration is higher and the collimation of light is more effective. Related to the ultraviolet detector but much more sensitive and selective is the fluorescence detector which has been shown to be feasible with SFC.

Spectrometric detection is also practiced with SFC. Traditionally, mass spectroscopy has been most studied with o.t. columns, however, packed column SFC/MS is gaining in popularity.[31] A variety of ionization sources have been used,but chemical ionization is the most common. Progress is rapidly being made with common HPLC/MS interfaces such as thermospray and particle beam for use with packed columns. Whatever the interface the following requirements must be met in order to be successful: (a) sufficient vacuum to handle mobile phase decompressed

Conclusion

Supercritical fluids possess many of the attributes necessary for HPLC, including low mobile phase viscosity, high analyte diffusivity, and good solubility for a wide range of analytes. More importantly, by changing the density of the mobile phase with a change in temperature and/or pressure, the observed chromatographic characteristics in an SFC separation can be changed. Thus a single, supercritical mobile phase can be used to afford a wide variety of separations without the time-consuming column equilibration necessary in high-performance liquid chromatography when changing mobile phase composition. Carbon dioxide is by far the most common mobile phase used in SFC. Polar compounds are widespread and are often nonvolatile and thermally labile. These properties make analysis by GC impossible without derivatization, and make method development for LC complex. SFC does not require solutes to be volatile, because separations are carried out at low temperatures, and method development is straightforward. However, supercritical carbon dioxide is nonpolar and therefore will not solvate many compounds of interest. The poor solvating powers of supercritical CO_2 for polar compounds necessitates the addition of a mobile phase modifier, such as methanol. The future of SFC at this time appears to be bright. The focus of applications and research will be traditional packed columns, modified CO_2 mobile phases, and numerous detectors which will be element-specific, flame-pharmaceuticals, chemicals, polymers, and natural products will be made. A sample of such applications follows.

References

1.   Klesper, E.; Corwin, A. H. and Turner, D. A.; J. Org. Chem., **27**, 700 (1962).

2. Sie, S. T.; van Beersum, W. and Rijnders, G. W. A.; Sep. Sci., **1**, 459 (1966).
3. Jentoft, R. E. and Gouw, T. H.; J. Chromatogr. Sci., **8**, 138 (1970).
4. Novotny, M.; Springston, S. R.; Peaden, P. A.; Fjeldsted, J. C. and Lee, M. L.; Anal. Chem., **53**, 407A (1981).
5. Lee, M. L.; Markides, K. E.; *Analytical Supercritical Fluid Chromatography and Extraction,* Provo, Utah: Chromatography Conferences, Inc. (1990).
6. Berger, T. A.; Packed Column Supercritical Fluid Chromatography, Royal Society of London (1996).
7. Gere, D. R.; Science, **222**, 253 (1993).
8. Strode, J. T. B., Taylor, L. T., Howard, A. L., Ip, D., and Brooks, M. A., J. Pharm. Biomed. Anal., **12**, 1003 (1994).
9. Schwartz, H. E.; Barthel, P. J.; Moring, S. E. and Lauer, H. H.; LC/GC, **5**, 490 (1987).
10. King, J. W.; Hill, H. H. and Lee, M. L.; "Physical Methods of Chemistry Series, eds. Rossiter, B. W. and Baetzold, R. C., John Wiley & Sons, Inc., New York (1993).
11. Reid, R. C.; Prausnitz, J. M.; Sherwood, T. K.; *The Properties of Gases and Liquids*, 3rd ed., New York: McGraw-Hill (1977).
12. Lauer, H. H.; McMannigill, D. and Board, R. D.; Anal. Chem., **55**, 1370 (1983).
13. Berger, T. A. and Deye, J. F.; Anal. Chem., **62**, 1181 (1990).
14. Berger, T. A. and Deye, J. F.; J. Chromatogr., **547**, 377 (1991).
15. Berger, T. A. and Deye, J. F.; J. Chromatogr. Sci., **29**, 26 (1991).
16. Schweighardt, F. K. and Mathias, P. M.; J. Chromatogr. Sci, **31**, 207 (1993).
17. Strode, J. T. B., Leichter, E., Taylor, L. T., and Schweighardt, F. K., Anal. Chem., **68**, 894 (1996).
18. Levy, J. M. and Guzowski, J. P.,; J. Chromatogr. Sci., **26**, 194 (1988).
19. Chester, T. L. and Innis, D. P.; "Quantitative Aspects of Capillary SFC", Abstracts of 5th International Symposium on SFE/SFC, Baltimore, MD, January, 1994, p. 7.
20. Markides, K. E.; Fields, S. M. and Lee, M. L.; J. Chromatogr. Sci., **24**, 254 (1986).
21. Chang, H-C. K.; Markides, K. E.; Bradshaw, J. S. and Lee, M. L.; J. Chromatogr. Sci., **26**, 280 (1988).
22. Ashraf-Khorassani, M.; Taylor, L. T. and Henry, R. A.; Anal. Chem., **60**, 1529 (1989).
23. Macaudiere, P.; Caude, M.; Rosset, R. and Tambute, A.; J. Chromatogr. Sci., **27**, 583 (1989).
24. Macaudiere, P.; Caude, M.; Rosset, R. and Tambute, A.; J. Chromatogr.; **405**, 135 (1987).
25. Li, W.; Malik, A.; Lee, M. L.; J. Microcol. Sep., **6**, 557 (1994).
26. Shen, Y.; Li, W.; Malik, A.; Reese, S. L.; Rossiter, B. E.; Lee, M. L.; J. Microcol. Sep., **7**, 411 (1995).
27. Francis, E. S.; Lee, M. L.; Richter, B. E.; J. Microcol. Sep., **6**, 449 (1994).
28. Ibanez, E.; Tabera, J.; Reglero, G.; Herraiz, M.; J. Agric. Food Chem., **43**, 2667 (1995).
29. Richter, B. E.; Bornhop, D. J.; Swanson, J. T.; Wangsqaard, J. G. and Anderson, M. R.; J. Chromatogr. Sci., **27**, 303 (1989).
30. Fields, S. M.; Markides, K. E. and Lee, M. L.; Anal. Chem., **60**, 802 (1988).
31. Pinkston, J. D.; Owens, G. D.; Burkes, L. J.; Delaney, T. E.; Millington, D. S. and Maltby, D. A.; Anal. Chem., **60**, 962 (1988).

Chapter 12

Analysis of Natural Products: A Review of Chromatographic Techniques

Christina Borch-Jensen and Jørgen Mollerup

Department of Chemical Engineering, Building 229, Technical University of Denmark, 2800 Lyngby, Denmark

The analysis of natural products like fats, seeds, oils, and tissues is compared for gas chromatography (normal and high temperature) and packed and capillary SFC. Examples are given by the determination of fatty acid composition and the analysis of triglycerides. The fatty acid determination is compared for the oils of cape marigold (*Dimorphoteca pluvialis*) a hydroxy fatty acid containing oil and a polyunsaturated fish oil from the sand eel (*Ammodytes sp.*). The triglyceride analyses of samples of butter fat and raw fish oil from the sand eel are compared for SFC and GC. In fatty acid determination and triglyceride analysis it is shown to be of great importance to consider the nature of the sample before selecting chromatographic technique. The presence of functional groups like hydroxy or epoxy groups in the fatty acids can affect the choice of technique and so can the chain length and degree of unsaturation of the fatty acid moieties. Another important aspect to consider is the purpose of the analysis. Process control and fingerprint analyses often do not require a total separation of all components in the sample, which means that the analysis time can be shortened considerably compared to an analysis where total quantification of all components is the goal.

In the analysis of fats, oils, seeds and tissues, there are many different ways to proceed. In the selection of chromatographic technique there are several possibilities, including gas (GC), liquid (LC), supercritical fluid chromatography (SFC), capillary electrophoresis (CE) and thin-layer chromatography (TLC). The time allowed for method development is often short and therefore the standard methods from IUPAC or AOCS, which provides methods for different types of analyses, are very popular.

When selecting a technique for one specific analytic problem there are some important aspects to consider. The selection of technique is often strongly dependent of the equipment available in the laboratory. If a standard method prescribes analysis by one specific technique, the question is if one should buy new equipment to fulfill the requirements of the standard method or if it would be sufficient to develop a new method running on the old equipment. Most standard methods in the fats and oils applications uses LC and GC. Another very important aspect to consider is how many samples are to be run by the method. If the method is used for routine analyses, the shortest, cheapest and most easy sample preparation and analysis is required, while if it is just a matter of few samples for instance for research work, this is of less importance. It would also be important to start by defining the goal of the analysis at the beginning of the process. If the sample is for instance palm oil, there could be several results of the analysis, like quantification of tocopherols or identification of triglycerides. In many cases the different results would require different analysis techniques and one method would not necessarily give both results. In this paper we will compare GC and packed and capillary SFC for different analytical problems related to natural samples.

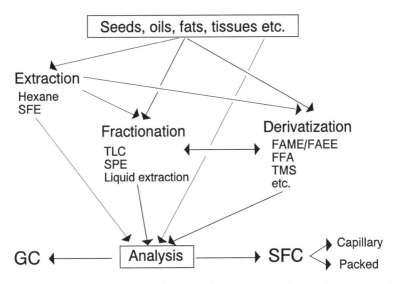

Figure 1   The numerous path ways from raw sample to chromatographic analysis for natural products.

Figure 1 illustrates different pathways possible in the analysis and sample preparation when dealing with natural samples. If the start material is seed or tissue the first step is normally an extraction, which can be done supercritical by SFE or by the conventional organic solvent extraction. After extraction one can choose to fractionate, derivatize or to analyze directly. Fractionation gives more choices, TLC, solid-phase extraction (SPE) or liquid extraction. Derivatives can be made as methyl or ethyl esters of fatty acids, as free fatty acids and there are the possibilities to mask functional groups like hydroxy groups with for instance silylating agents. The final choice is concerning analysis technique. Once the analyte has been prepared there is one more choice: GC, packed or capillary SFC?

The following examples are determination of fatty acid composition and analysis of triglycerides. With these examples, we wish to show how molecular weight, functional groups, degree of saturation can affect the choice of analytical technique.

Determination of fatty acid composition.

The determination of fatty acid composition normally requires some kind of derivatization. The most used type of derivatives are fatty acid methyl or ethyl esters (FAME/FAEE), less used are free fatty acid (FFA) derivatives. Table I illustrates which technique can be used under which circumstances. The fatty acids have been divided in three different types, "normal", hydroxy and epoxy. the first type include most animal, fish and vegetable oils and the most common way to proceed when dealing with these samples are GC of FAMEs or FAEEs, although both packed (*1,2*) and capillary (*3,4*) SFC have been applied. There are certain seed oils, which have rather high contents of fatty acids with functional groups like hydroxy or epoxy. The reason why these too groups of functional groups can not be treated as one is due to the fact that these groups have different properties. The epoxy group is very reactive, but not thermolabile. For this reason the analysis by GC is not a problem, but the derivatization reaction is not straight forward, because the epoxy groups are affected by the acidic reagents normally

Table I.   Determination of fatty acid composition of different types of fatty acids and different types of derivatives

Type of derivative	Nature of fatty acid		
	"Normal"	Epoxy	Hydroxy
FAME/FAEE	GC	(GC)	GC(TMS)
FFA	GC	GC/SFC	SFC/GC(TMS)

used in FAME/FAEE preparation. Basic reaction is possible, but this method will not derivatize the free fatty acids present in the seed oil in the first place. This is why the most safe way to make derivatives of epoxy fatty acid containing triglycerides are preparation of FFA derivatives (*5*) which can readily be analyzed by capillary SFC or GC on a polar column. The hydroxy group can be thermolabile, but this is strongly dependent on the mutual positions of the hydroxy group and the double bonds in the fatty acid chain. Fortunately it is easy to mask the hydroxy groups by trimethyl silyl groups (TMS derivatives) and analyze the samples by GC. This method will give both the fatty acid composition of the normal fatty acids and the content of the hydroxy fatty acids. However, the TMS reagent is rather expensive and the sample preparation time is long. An alternative method is SFC of FFA derivatives like with the epoxy fatty acid containing oils. This does not require TMS reagent and the sample preparation is quite fast.

The analysis of fish oil FAMEs by GC is a good example of a standard method which is well working and not easily replaced by other techniques (*4*). Figure 2 shows the GC-MS chromatogram from such an analysis. Almost 70 peaks are resolved in less than 40 minutes. The method for the derivatization and analysis is a standard AOCS method (*6*). Applying capillary SFC for the same type of sample will give the chromatogram shown in Figure 3, a polar SFC column separation of fish oil FAMEs. The retention time is three times as long as in the GC analysis and the number of components separated are decreased to about 35. Applying a packed, polar SFC column, the chromatogram will be as in Figure 4. The analysis time is much shorter, but the separation is insufficient, which is mainly due to the short 25 cm column.

In the case of hydroxy fatty acids, the analysis by both SFC and GC have disadvantages and advantages. The method using SFC of FFA has the advantage of an easy sample preparation, while the chromatographic analysis takes 40 minutes and still the separation of the normal fatty acids are not satisfactory (*7*). The SFC chromatogram is shown in Figure 5. By GC of TMS-FAMEs, Figure 6, the chromatographic separation of both normal and hydroxy fatty acids is very good with a total analysis time of only 22 minutes. Unfortunately, the method is expensive. In cases like this, it is important to decide what one wants to obtain, because the SFC method is a cheap and easy way to determine the total content of hydroxy fatty acid, while the GC method is more expensive, but on the other hand gives the distribution of both normal and hydroxy FAMEs in a short analysis time.

Analysis of triglycerides.

SFC has often been stated as an excellent technique for the analysis of the high molecular weight triglycerides and a variety of different triglycerides have been analyzed (*8-13*) by this technique. As can be seen from Table II, the suitability of SFC depends on both the chain length of the fatty acids in the molecules and on whether they are saturated or unsaturated. In the case of short to medium chain triglycerides with few doublebonds high temperature GC (HT-GC) often offers a separation that is better than in SFC and within a much shorter analysis time. In HT-GC the maximum allowed column temperature is in the range of 380 to 400°C. The high temperatures are required to elute triglycerides with up to 54 carbon atoms in the fatty acid chains. If however, the triglyceride contains any polyunsaturated fatty acids, these will not be eluted safe from the column and if the total number of carbons in the triglyceride chains is much higher than 54, the triglyceride will not be eluted at all. In such cases SFC is the best choice.

Table II. Analysis of triglycerides with different chain length and degree of unsaturation

	Degree of unsaturation	
Chain Length	Saturated	Unsaturated
Short to medium	GC	SFC/GC
Medium to long	GC/SFC	SFC

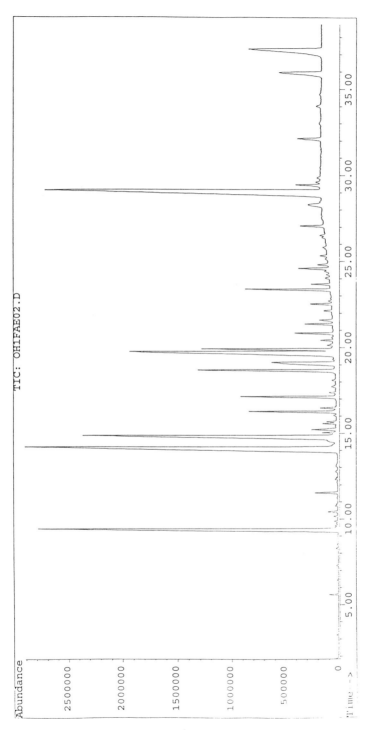

Figure 2. Analysis by GC of fish oil fatty acid methyl esters. Column: HP-FFAP, 25 m, 0.2 mm. Flow: 1 mL/min. Carrier gas: He. Temperature program: From 140°C (1 min) to 220°C with 3 °C/min, then 220°C for 10 minutes. Detector: MSD.

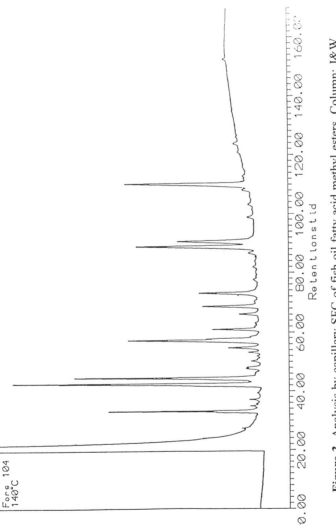

Figure 3. Analysis by capillary SFC of fish oil fatty acid methyl esters. Column: J&W DB-225, 20 m, 0.1 mm. Flow: 1 mL/min. Mobile phase: CO_2. Temperature: 140°C. Density program: From 0.15 g/mL (25 min) to 0.225 g/mL with 0.001 g/mL/min, then to 0.355 g/mL with 0.002 g/ml/min, then 0.355 g/mL for 30 minutes. Detector: FID.

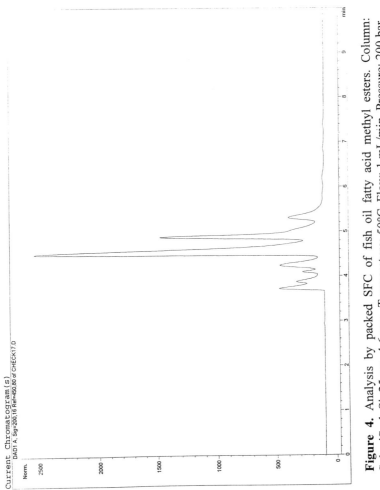

Figure 4. Analysis by packed SFC of fish oil fatty acid methyl esters. Column: SpheriSorb Si, 25 cm, 4.6 mm. Temperature: 50°C. Flow: 1 mL/min. Pressure: 200 bar. Mobile phase: 5 mol% methanol in CO_2. Detector: UV at 200 nm.

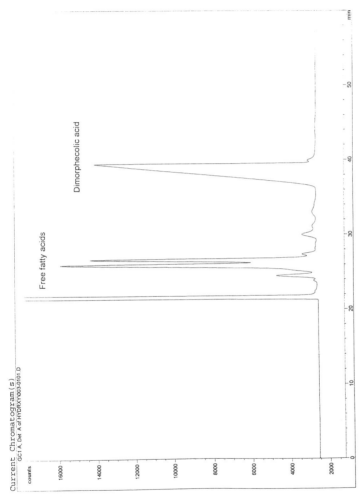

Figure 5. Analysis by capillary SFC of hydroxy free fatty acids. Column: J&W DB-225, 20 m, 0.1 mm. Flow: 1 mL/min. Mobile phase: CO_2. Temperature: 140°C. Density program: From 0.2 g/mL to 0.6 g/mL with 0.001 g/mL/min, then 0.6 g/mL for 20 minutes. Detector: FID.

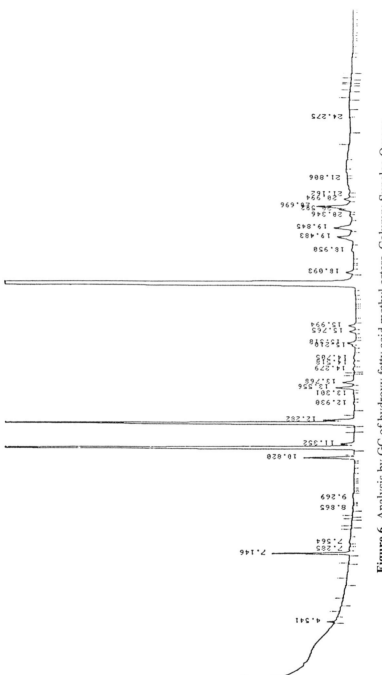

Figure 6. Analysis by GC of hydroxy fatty acid methyl esters. Column: Supelco Omega Wax, 25 m, 0.2 mm. Flow: 1 mL/min. Carrier gas: He. Temperature program: From 140°C (1 min) to 220°C with 3 °C/min. Detector: FID.

The analysis of butter fat triglycerides is compared for HT-GC and capillary SFC in Figures 7 and 8. In both cases a non-polar column was used to achieve a separation according to carbon number. The analysis time for HT-GC is half of that for SFC and the separation is also better by this method, which makes HT-GC the best choice in this case.

Figures 9 to 12 compare the analysis of raw fish oil by HT-GC, packed and capillary SFC. The HT-GC chromatogram, Figure 9, shows what happens when exposing high molecular weight polyunsaturated triglycerides to 370°C. Although the analysis time is less than 30 minutes a quantification is not possible because the late eluting components are not separated and because one can not be sure that all components are eluted from the HT-GC column. Applying capillary SFC and a non-polar column, Figure 10, will give a total separation of the fish oil according to carbon number. The analysis time is 120 minutes, but this analysis allows a total quantification of free fatty acids, free cholesterol, wax esters, di- and triglycerides in one chromatographic run. The analysis shown in Figure 10 is however an example of a less complicated marine oil sample. Depending on which lipid groups are present in the oil, hydrogenation or fractionation by TLC might be necessary to obtain a total quantification (*5*). With a packed ODS SFC column, Figure 11, and a methanol modified mobile phase the fish oil is separated according to the polarity of the components, which gives too many peaks for baseline separation. However, the method gives a fast determination of the free fatty acid content, thus being suited for process control in a deodorization process. In this case the analysis time could be shortened considerable, because no separation of the triglyceride species is needed. The separation of fish oil triglycerides by silver nitrate or argentation chromatography is shown in Figure 12 (*13*). In argentation chromatography, the components are separated solely according to number of doublebonds with the smallest number of doublebonds eluting first. In Figure 12, the numbers refer to the total numbers of doublebonds in the triglycerides. Although the separation is not complete, this type of chromatogram offers a very good fingerprint of the fish oil.

The Figures 9 through 12 have shown that the analysis of polyunsaturated long chain triglycerides is best done by SFC. The choice between packed and capillary is in this case however a question of which kind of result is needed. The separation needed for process control, total quantification or fingerprint analyses is not the same and it is important to choose the right technique and method.

Conclusions.

In conclusion, this paper has shown a few examples of the many possible ways to proceed from sample to resulting GC or SFC chromatogram. It has been shown that the selection of technique is strongly dependent of the specific sample. Determination of fatty acid composition can be done by both GC and SFC, depending on whether the fatty acids are normal, hydroxy or epoxy fatty acids. Triglyceride analysis by HT-GC is restricted to low to medium molecular weight components with no or only few double bonds, because the high temperatures applied in this technique would destroy any poly-unsaturated triglycerides. Triglyceride analysis by SFC can be done by both packed and capillary columns. The capillary column analysis offers quantification of free fatty acids, free cholesterol, wax esters, di-and triglycerides, while packed column SFC of fish oil triglycerides gives the possibilities of process control and finger print analysis.

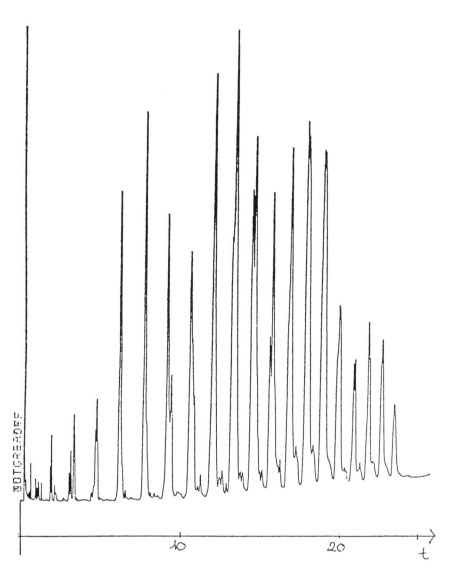

Figure 7. Analysis by HT-GC of butter fat triglycerides. Column: Chrompack HT-SimDist, 10 m, 0.3 mm. Flow: 13 mL/min. Carrier gas: He. Temperature program: From 200° to 370°C with 10 °C/min, then 370°C for 10 minutes. Detector: FID.

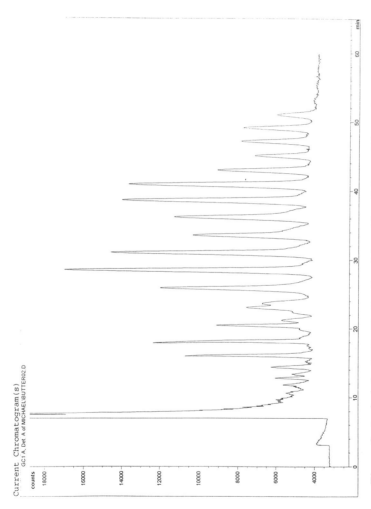

Figure 8. Analysis by SFC of butter fat triglycerides. Column: Dionex SB-Methyl, 10 m, 0.05 mm. Flow: 1 mL/min. Mobile phase: CO_2. Temperature: 170°C. Density program: From 0.3 g/mL to 0.36 g/mL with 0.001 g/mL/min. Detector: FID.

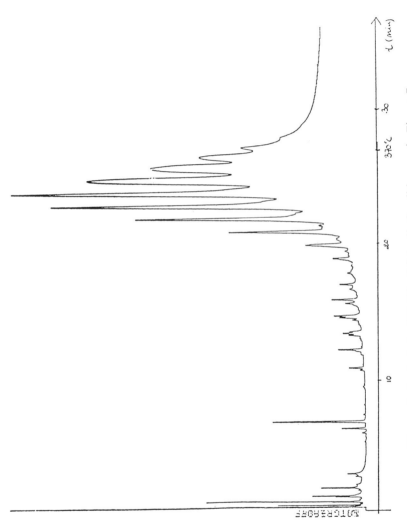

Figure 9. Analysis by HT-GC of fish oil. Conditions as in Figure 7.

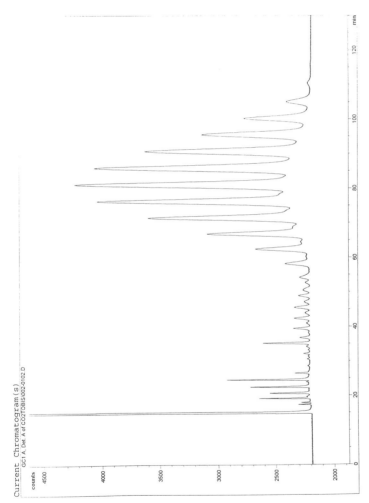

Figure 10. Analysis by capillary SFC of fish oil. Column: J&W DB-5, 20 m, 0.1 mm. Flow: 1 mL/min. Mobile phase: CO_2. Temperature: 170°C. Density program: From 0.3 g/mL to 0.452 g/mL with 0.004 g/mL/min, then 0.452 g/mL for 14 minutes, then to 0.52 g/mL with 0.001 g/mL/min, then 0.52 g/mL for 10 minutes. Detector: FID.

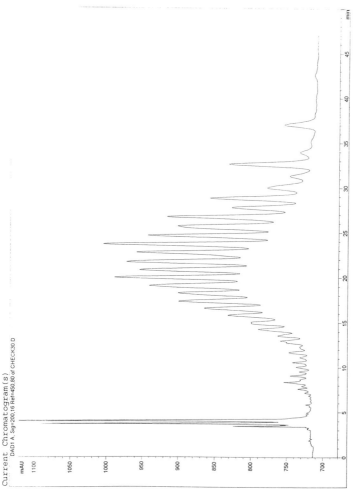

Figure 11. Analysis by packed SFC of fish oil. Column: SpheriSorb ODS, 25 cm, 4.6 mm. Temperature: 65°C. Flow: 1.5 mL/min. Pressure program: 200 bar for 15 minutes, then to 225 bar with 1 bar/min. Mobile phase: 5 mol% methanol in CO_2. Detector: UV at 200 nm.

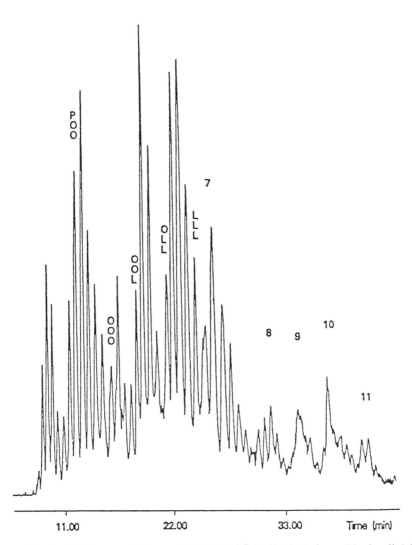

Figure 12. Analysis by argentation packed SFC of fish oil (*6*). Column: Nucleosil 4 SA impregnated with AgNO₃, 15 cm, 0.7 mm. Temperature: 100°C. Pressure: 340 atm. Mobile phase: CO_2. Modifier: Acetonitrile:2-Propanol, 9:1 (mol). Modifier gradient: From 1 to 10 mol % modifier in CO_2. Detector: ELSD.

References.

(1)   Sakaki, K. *J. Chrom.*, **1993**, *648*, 451-457.
(2)   Görner, T, Perrut, M. *LC GC International*, **1989**, *2(7)*, 36-39.
(3)   Demirbüker, M., Hägglund, I., Blomberg, L.G., *J. Chrom.*, **1992**, *605*, 263-267.
(4)   Staby, A.; Borch-Jensen, C.; Jensen, B.; Mollerup, J. *J. Chrom.*, **1993**, *648*, 221-232.
(5)   Borch-Jensen, C.; Mollerup, J. *JAOCS*, accepted for publication.
(6)   American Oil Chemists Society, AOCS Official Method Ce-2-66, AOCS Press, Champaign, IL, **1989**.
(7)   Borch-Jensen, C.; Jensen, B.; Mathiesen, K.; Mollerup, J. *JAOCS*, submitted.
(8)   Manninen, P., Laakso. P., Kallio, H. *JAOCS*, **1995**, *72(9)*, 1001-1008.
(9)   Baiocchi, C., Saini, G., Cocito, C., Giacosa, D., Roggero, M.A., Marengo, E., Favale, M. *Chromatographia*, **1993**, *37(9/10)*, 525-533.
(10)   King, J.W., *J. Chom. Sci.*, **1990**, *28*, 9-14.
(11)   Houpalathi, R., Laakso, P., Saaristo, J., Linko, R., Kallio, H., *HRC*, **1988**, *11*, 899-901.
(12)   Borch-Jensen, C.; Mollerup, J. *Chromatographia*, **1996**, *42(5/6)*, 252-258.
(13)   Blomberg, L.; Demirbüker, M.; Andersson, P.E.; *JAOCS*, **1993**, *70(10)*, 939-946.

Chapter 13

The Influence of Entrainers on the Extraction of Polycyclic Aromatic Hydrocarbons from Contaminated Soils

A. Schleussinger, I. Reiss, and S. Schulz

Institute of Thermodynamics, University of Dortmund, Emil Figge Strasse 70, D–44221 Dortmund, Germany

The extraction of polycyclic aromatic hydrocarbons (PAH) with supercritical carbon dioxide from contaminated soils can be improved by the use of entrainer. In comparison to alcohols or alkanes, water yields good results as moisture of the soil due to its low solubility in the supercritical fluid. Thus, water is only slowly driven out of the soil and can therefore influence the adsorption of the contaminants on the soil particles even at longer extraction times. The agglomeration of soil with elevated water content results in an optimum humidity for the extraction of contaminated soil in the range of 8 - 15 %.

Supercritical fluids (SCF) have several exceptional properties making them to interesting solvents for chemical engineering applications. Slight changes in temperature or pressure can cause large changes in the density of these fluids and consequently in their solubility power. Supercritical fluid densities compare to liquid densities indicating their ability to solubilize heavy nonvolatile organics. Their viscosities and diffusivities are intermediate to typical liquid and gas values of these properties. Thus the extraction with supercritical fluids (SFE) is a promising technique for a rapid and effective cleanup of soils, sediments and other wastes that are contaminated with hazardous organics e.g. grinding and metal cutting waste (*1*). In contrast to liquid solvents, the mass transfer characteristics of SCF let them easily penetrate fine grained matrices such as soil consisting of particles with a wide range of diameters. Because of this, the SFE process is common for the preparation of environmental samples. A number of experimental and theoretical studies have recently been published showing the applicability of the SFE process for the cleanup of hydrocarbon contaminated soils (*2,3*). This technique is also discussed for the extraction of organics from waste water (*4*).

Carbon dioxide is mostly used as solvent because it is virtually inert, cheap, has moderate critical properties and leaves no solvent residue on the processed soil.

A lot of experiences are already made with carbon dioxide as solvent and the purchase of apparatus for this solvent can easily be done, therefore it is also used for our application of the SFE process.

Besides the extraction, the regeneration of the loaded solvent has also to be considered for the design of a SFE process. Several processes can be used after the extraction for this purpose (5). Aside from the adsorption on activated carbon, lowering the solubility for organics can easily be done for several orders of magnitude by a pressure reduction which leads to precipitation of the contaminants (6). In contrast to an isobaric process, the recirculation of the solvent requires a high amount of energy and therefore, the soil treatment costs are strongly related to the necessary amount of solvent for the cleanup. Due to the high pressure of the SFE process, all possibilities to lower the solvent consumption should be investigated (7). One possibility to improve the extraction efficiency is the addition of cosolvents to the supercritical fluid. The addition of an other substance modifies the properties of the solvent and allows an adjustment for an optimum separation. To work out the effect of these entrainers, spiked soil and soil from former gasplant sites were extracted. For first theoretical investigations the extraction of spiked model matrices are easier to interpret because of the fewer components and less interactions between them.

Our research work focuses on the removal of polycyclic aromatic hydrocarbons because of their low volatility, which makes them the most difficult components to be removed from contaminated soil. Perylene was chosen as model substance, because its vapor pressure is an order of magnitude lower than the sublimation pressure of benzo(a)pyrene which is of major interest due to its cancer potential. Thus, perylene is one of the PAH with the lowest solubility, but it is less toxic. Some properties of these PAH are summarized in table I. The physical properties of SCF allow the extraction of fine grained matter, therefore silt with an averaged particle size of 30μm was used as matrix for the extraction of perylene.

Table I. Properties of some polycyclic aromatic hydrocarbons

PAH	structure	melting point [°C]	boiling point [°C]	vapor pressure at 100 °C [bar]
pyrene		156	393	$1.80 \cdot 10^{-5}$
chrysene		256	441	$1.58 \cdot 10^{-7}$
perylene		578	497[a]	$4.03 \cdot 10^{-9}$
benzo(a)pyrene		177	496	$1.45 \cdot 10^{-7}$

[a]predicted (White, C.M. *J. Chem. Eng. Data* **1986**, 2, pp 198)

The influence of entrainer on the supercritical extraction of contaminated soils can be considered under three major aspects: selectivity, solubility and adsorption. These aspects will be discussed in the following with theoretical and experimental results.

Experimental Facilities

Two extraction plants were applied in this study. One lab-scale plant with a 30 ml-extractor was used for initial experiments allowing a solvent mass flow of 0.13 kg/h. A schematic of the other plant with a 4 l-extractor and a solvent cycle is given in figure 1. The CO_2 from the storage vessel (V) is subcooled in a heat exchanger (H1) and pumped to extraction pressure by the pump (P1). A mass flow up to 40 kg/h can be established. In the heat exchanger (H2) the CO_2 reaches extraction temperature, before it enters the extraction autoclave (E) with the contaminated soil. The temperature of the CO_2 can be changed in the heat exchanger (H3) after passing the extractor for adjusting the conditions before expansion in the separation autoclave (S) where the contaminants precipitate. The regenerated clean CO_2 is lead back to the storage vessel for further use. A small diaphragm pump (P2) allows the feeding of entrainer to the solvent. Both plants are equipped with UV-detectors for measuring the concentration of the contaminants in the solvent. For this, a part of the solvent is branched off and lead through a heated capillary to a HPLC-UV-detector (Jasco UV-970) with a high pressure cell. This allows an easier determination of the extraction progress with only one experiment. Thus, a deeper investigation of the process and its limiting mechanisms is possible. Figure 2 shows the data obtained by the UV-detector for the extraction of perylene. The course of the signal over the wavelength is characteristic for each substance. The course of the signal over time can be taken as measure for the concentration. The GC analysis of the soil before and after the treatment allows to calculate the course of concentration from the detector signal by means of a mass balance. The integration of the concentration allows the determination of the concentration in the soil as a function of time. A UV-detector provides also useful information for the modeling of the extraction process such as residence time distribution (*8*).

Experimental Results

For better comparison of the results, it is convenient to display the course of extraction in a dimensionless form. A good measure for the quality of the removal is the extraction yield ξ, the ratio of extracted mass Δm of contaminant to the initial mass m_0:

$$\xi = \frac{\Delta m_{PAH}}{m_{PAH,0}} = \frac{q_0 - q(t)}{q_0}.$$

The extraction yield can also be defined by using the initial and the final concentration in the soil. A value of one stands for a complete removal. A dimensional analysis of the SFE process delivers the specific solvent ratio Λ, the mass of solvent over the mass of soil, as a second important value for the scale-up:

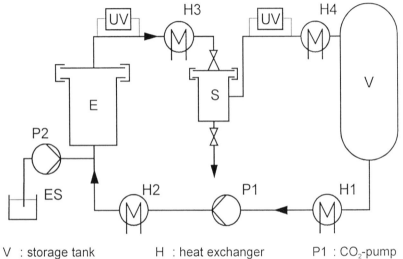

V : storage tank   H : heat exchanger   P1 : CO$_2$-pump
E : extraction autoclave S : separation autoclave UV : UV-detector
P2 : entrainer pump   ES: entrainer stock

Figure 1. Schematic of the pilot scale extraction plant with solvent cycle.

$$\Lambda = \frac{m_{CO_2}}{m_{soil}} = \frac{\dot{m}_{CO_2} \cdot t_{ex}}{m_{soil}} .$$

The specific solvent ratio can also be interpreted as dimensionless extraction time. Therefore, the results of the extraction experiments are shown as extraction yield against specific solvent ratio.

Spectroscopic Measurements. The advantage of analyzing the extraction process with spectroscopic methods can be seen in figure 3. It shows the extraction of pyrene and perylene as well as their simultaneous extraction from silt. The endpoints of each line are determined by GC-analysis of the extracted soil samples. The course of extraction up to the endpoint is calculated from the UV-detector signal. In case of a multicomponent extraction, each substance can be identified by its spectrum. This allows the determination of the contribution from each substance to the UV-detector signal and the analysis is possible as well. As first result it can be seen that pyrene and perylene do not effect each other as this is known from other substances e.g. phenantrene and benzoic acid (9). These curves also show that the extraction can be divided into a first part with a high slope indicating a rather high concentration and a high extraction rate and a second part where the extraction progress ceases slowly. In this part, the extraction yield approaches a limit which can not be overcome even with much more solvent. This shape of the curve can be explained by assuming a mass transfer limitation for the first part and a limitation by adsorption for the second part. Therefore, a solubility enhancement by entrainer can only affect during the first part whereas the influence on adsorption will alter the extraction progress in the second part. The third aspect of entrainer, their selectivity is not as important for soil decontamination purpose because all contaminants should be extracted without any distinction.

It is known from analytical applications that methanol is an efficient entrainer for the removal of PAH from environmental samples (10). Polar and nonpolar entrainers have to be distinguished because of their different behavior. Therefore, we used n-hexane as a nonpolar entrainer and methanol, ethanol and water as polar entrainers in our experiments. Water is found in rather all soils as natural moisture and because of this its effect on the technical SFE-process is interesting. Another important aspect is the distribution of the entrainer in the extraction vessel, the plant and the feeding method, discontinuously to the soil or continuously to the SCF. Apart from the addition of entrainer, even the use of technical carbon dioxide results in higher extraction yields than the extraction with pure carbon dioxide.

Effect on Solubility. The solubility enhancement of solvents like alcohols or acetone for PAH in supercritical fluids has already been measured for several PAH (11,12). Due to the nature of our extraction plants, solubility measurements are not possible, because no equilibrium can be achieved. Only the maximum concentration during the extraction might be taken as a measure for the solubility because of mass transfer resistance in the extractor. The extraction of glass beads spiked with perylene to high concentrations should only slightly be influenced by adsorption and mass transfer resistance. But the results showed no great differences in the maximum concentration when extracted with and without entrainer. The results were verified

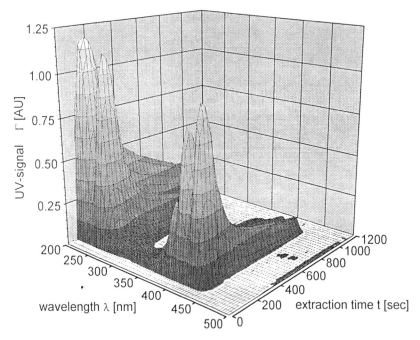

Figure 2. UV-detector signal of an extraction of perylene.

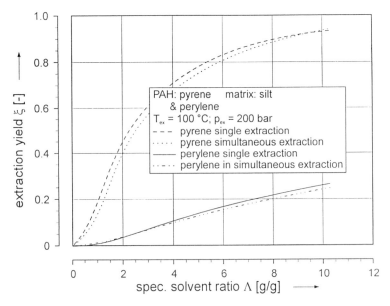

Figure 3. Comparison of single and simultaneous extraction
of pyrene and perylene.

theoretically by calculating the entrainer effect of water on the solubility of PAH. This was done for pyrene because all necessary date are available from the literature. The binary mixture of water and CO_2 is calculated with the Peng-Robinson-Equation of state and a mixing rule proposed by Panagiotopoulos and Reid (*13*). The description of the properties of pure water has to be improved by adjusting the parameters of the equation of state to experimental data as proposed in (*14*). This attempt provides a good description of the binary mixture. Results for the solubility calculations of several PAH and benzene show that the interaction parameter are virtually the same for these substances. For the calculation of the interaction parameter for pyrene and water, no data were available where the use of EOS is possible. Therefore, it was estimated from benzene-water LLE and VLE data and benzene-water-CO_2 distribution equilibrium data (*15*). The result of this calculation is shown in figure 4. The entrainer effect s is calculated as the ratio of the solubility enhancement in the ternary mixture y_{ter} to the solubility in pure CO_2 y_{bin}

$$s = \frac{y_{ter} - y_{bin}}{y_{bin}}.$$

Although pyrene is hardly soluble in water, the solubility of pyrene in moistened CO_2 is even higher. This solubility enhancement is a result of the higher density of the CO_2-water mixture because of the higher density of water. The same phenomenon was already measured for DDT which is as less soluble as pyrene in water (*16*). The experiments and the calculations indicate that the selected entrainer can only slightly alter the solubilities of the PAH in the SFE-process. However, the extraction of organic contaminants is hardly limited by the solubility.

Effect on Adsorption. For the required high extraction yield for soil decontamination purpose, the course of extraction at higher extraction times with its lower concentration of the contaminants in the fluid and the soil is more considerable. The adsorption seems to be the most important factor limiting the extraction yield. With increasing extraction time the stronger adsorbed molecules of the contaminants on the soil result in an increasingly difficult improvement of the cleanup.

The extraction results presented in the following show this effect of entrainers on the extraction. For this reason, mild extraction conditions are chosen to work out this effects and not for the achievement of high extraction yields. Otherwise, the temperature must be as high as possible.

Figure 5 shows the influence of several entrainers on the extraction of pyrene when fed to the soil. Even without entrainer pyrene can be removed under these conditions to more than 90 %. Therefore, the entrainers are leading only to a higher extraction rate at the beginning and to a slightly improved extraction yield. In contrast, the extraction of perylene is much more effected. Besides the higher extraction rate the extraction yields are increased from 50 % up to 80 % as shown in figure 6. Methanol accelerates the extraction most, followed by ethanol. The enhancement by water is weaker. This experiment shows a difference in the shape of the extraction curve between water and the alcohols. The curve with water resembles the curve without entrainer but is shifted to a higher extraction yield. Even at the end, the slope of the curve indicates further increase of the extraction yield. In

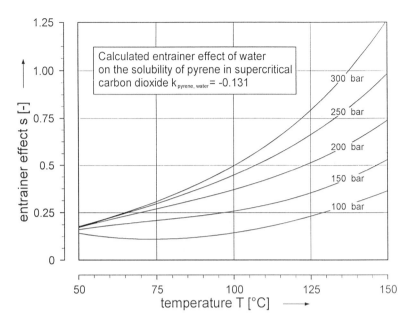

Figure 4. Calculated enhancement of the solubility of pyrene in supercritical carbon dioxide by water.

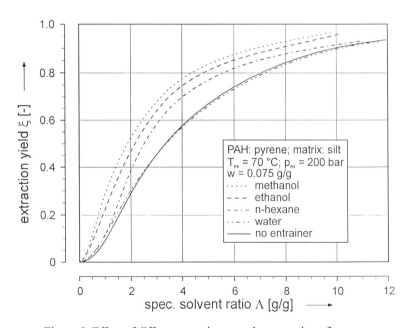

Figure 5. Effect of different entrainers on the extraction of pyrene from spiked silt.

contrast, the extraction curves with the alcohols have a higher slope at the beginning but at a spec. solvent ratio of around $\Lambda = 5$ g/g the curves are becoming more and more flat. The extraction stops at an extraction yield of 70 % or 80 % respectively. This difference is a result of the much higher solubility of the alcohols in comparison to water. At this conditions, the solubility of water is around 1.1 mol-% (*17*) and methanol is completely miscible. The alcohols are driven out of the extractor during the extraction with the contaminants and therefore they can not effect the cleanup in the second phase when the adsorption becomes more and more important. Water is only slightly soluble in supercritical carbon dioxide therefore it is only faintly extracted from the soil and it can improve the extraction at higher solvent ratios also. n-hexane showed only little effect on the adsorption and is therefore left out of the further investigations.

In figure 7 the continuos entrainer feeding to the SCF and the discontinuous feeding to the soil are compared. For convenience, methanol is used as entrainer instead of water because of its higher solubility. The feeding to the soil leads to the highest extraction rate at the beginning because the entrainer is already spread in the extractor volume and consequently the extraction yield is improved. In contrast, the continuous feeding results in a slower start but the effect on the adsorption at longer extraction times allows a higher extraction yield because of the described mechanism. As consequence there is an intersection of these two curves. The best result can be achieved with a combination of continuous and discontinuous feeding. A high extraction rate at the beginning is combined with an effective influence on the second part of the extraction. On the other hand it has to be noted that this method requires the highest amount of entrainer.

The limits of an improvement by the continuous entrainer feeding are shown in figure 8. Because of their solubility, the concentration of methanol and ethanol can be raised to a great extent. However, 10 mol-% of methanol or ethanol lead to the best extraction results. A higher concentration is of no use for the cleanup. As also seen in figure 6 methanol results in higher extraction yields than ethanol. This agrees with findings from analytical applications indicating the same concentration range for the best results (*9*). Higher concentrations are not hindering because these substances are miscible with carbon dioxide.

The extraction with an initial water content seems to be a very attractive method for the enhancement of the extraction because neither further apparatus nor auxiliary material are required. Figure 9 shows the extraction yield for perylene as a function of the initial moisture w of the silt. Only 3 % of moisture can increase the extraction yield from 35 % up to 80 % with a maximum extraction yield of approximately 90 % in the range of 4 to 9 % of initial moisture. But a water content of more than 10 % lowers the extraction yield. An explanation may be the agglomeration of the wet soil particles. These larger agglomerates lead to longer diffusion path and thereby to a worse cleanup. The optimum moisture depends on the nature of soil.

Extraction of Soil Samples. Finally, the results obtained with spiked materials are verified by the extraction of soil samples in the 4l-plant with a CO_2 mass flow of 20 kg/h. Due to the analysis with a gaschromatograph, benzo(a)fluoranthene and benzo(k)fluoranthene are determined together. Experiments with a fine grained soil sample with an averaged particle size of only 10

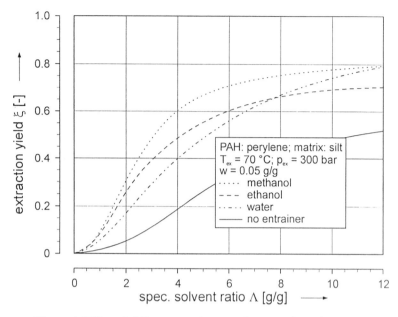

Figure 6. Effect of different entrainers on the extraction of perylene from spiked silt.

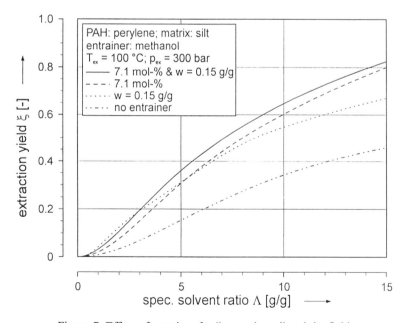

Figure 7. Effect of entrainer feeding to the soil and the fluid.

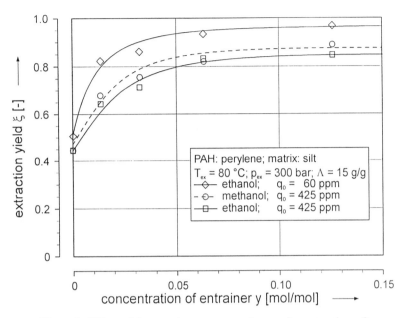

Figure 8. Effect of the entrainer concentration on the extraction of perylene from spiked silt.

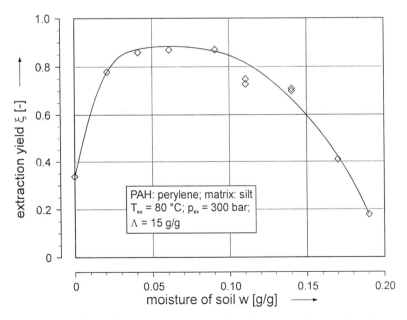

Figure 9. Effect of moisture on the extrction of perylene from spiked silt.

µm showed the best results at 12 % moisture. Figure 10 shows the influence of water on the extraction of the PAH chrysene from this material. These experiments were made at higher temperatures for a better cleanup with moistened and dry soil at 280 bar and 330 bar. As can be seen moisture is more useful than the higher extraction pressure. In both cases, the higher pressure improves the extraction only slightly. Moisture makes a large cleanup possible.

The same effect can be detected by extracting PAH from contaminated soils from former gasplant sites. As before, the optimum moisture is found at 12 %. The organic content is determined as weight loss by thermal gravimetric analysis. Figure 11 displays the extraction results for several PAH from sample with 5.4 w-% of organic matter. The extraction yields of the dry material are according to the volatility of the substances. The initial concentration of the analyzed PAH was 257 ppm. The extraction of the dry sample lowered the concentration to 145 ppm, according to an overall extraction yield of 45 %. The moistened sample is extracted a concentration of 15.4 ppm corresponding to an extraction yield of 94 %. As already seen for pyrene and perylene, the lower the extraction yield for dry soil, the higher is the improvement by water. This is not surprising, because the more volatile substances can be extracted without entrainer, whereas the adsorption binds the low volatile components stronger and water can have a greater effect. Unfortunately, this changes when the soil contains larger amounts of organic matter. Figure 12 shows the same comparison for a soil sample with 17.3 % of organic matter and an initial PAH concentration of 1133 ppm. As before, moisture leads to an improved cleanup but the remaining content is still to high for another use of this material.

Discussion

The calculations and experimental results are showing that the solubility is of minor interest for the extraction of organics from contaminated soil. This is a result of the mass transfer limitation and the influence of adsorption on the extraction process. Except for the beginning of the extraction, the concentrations in the solvent are far away from the solubility limit. Only entrainer concentrations up to roughly 10 mol-% are of interest. Otherwise the extraction conditions have a greater influence on the solubility than entrainer which can alter the solubility by a factor of around 2.

The dependence between the concentration of the soil q and the fluid phase concentration c can be described with an adsorption isotherm. The BET isotherm allows the representation of the solubility limit as well as the nonlinear shape at small concentrations. In this application, the concentrations are far away from solubility concentrations, even in the second phase of the extraction. Therefore, the more simple Freundlich isotherm is sufficient here (18):

$$q(c, T) = K_F(T) \cdot c^{\frac{1}{n}} .$$

The strong influence on the adsorption can be demonstrated when fitting the operating line of the extraction of perylene to that Freundlich isotherm as shown in figure 13. As already mentioned no equilibrium can be achieved in our extraction facilities. Therefore, the operating lines give an idea of the adsorption isotherms. The nonlinear shape confirms also additional measurements including the modeling of the extraction process (19) and results of other authors ($18,20$). The operating line

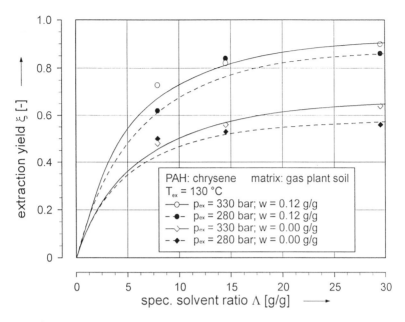

Figure 10. Extraction of chrysene from soil of a former gas plant site.

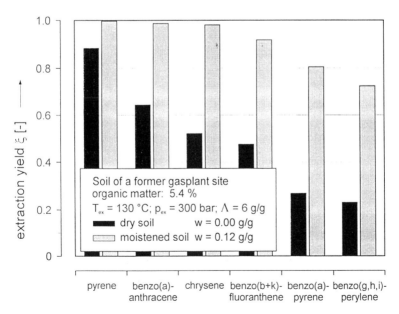

Figure 11. Extraction of a soil sample of a former gas plant site
with a moderate content of organic matter.

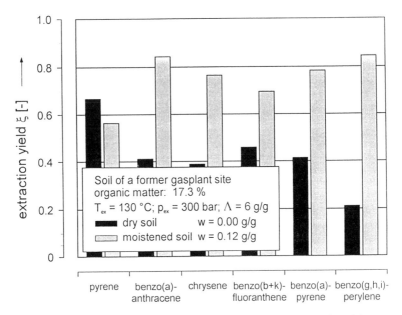

Figure 12. Extraction of a soil sample of a former gas plant site with a high content of organic matter.

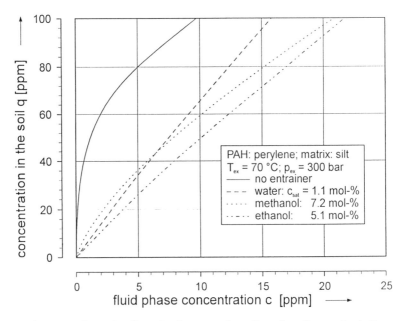

Figure 13. Operating lines for the extraction of perylene from spiked silt, fitted to Freundlich isotherms.

without entrainer ends at a concentration of about 10 ppm with a fluid concentration of virtually zero. A further reduction of this concentration is hardly possible because no contaminant can be driven out of the extractor by the solvent. In contrast, the operating lines of the experiments with entrainer are shifted to higher fluid concentrations allowing a nearly complete extraction because perylene is extracted even at a low soil concentrations. This change of the shape of the adsorption isotherm by moisture is measured for organics at normal conditions (*21*), too. In contrast to a linear adsorption isotherm this shape explains the negligible concentration of PAH in the solvent even after a long extraction time. This findings are not in contradiction to other investigators who found that moisture slows down the extraction rate (*22,23*). Their measurement were carried out with soil with a water content of 20 %. According to our results, this is not astonishing because the agglomeration hinders the cleanup more than the water can support it. As well, the dependence of extraction yield and organic content is a consequence of the adsorption limitation. In (*24*), the distinct relation between the partition coefficients of PAH and the organic content is shown for ambient conditions and in (*25*) for phenol and soil at extraction conditions.

An other important question for the application of this technique is the scale-up. Figure 14 shows the extraction yield for the second soil sample in the two different plants. The extraction time in the lab scale plant was ten times larger than in the technical plant according the mass flows to achieve the same specific solvent ratio. The extraction yields are nearly the same for all analyzed PAH indicating the great importance of this value.

Conclusion

The SFE of soil works in the pilot plant scale as well as in the analytical scale. The cleanup can significantly be accelerated by entrainer. This effect is attributed to ad-sorptive displacement and prevented readsorption of the contaminants by polar en-trainers. Also the better results with moistened soil indicates that adsorption and not the diffusivity is the limiting mechanism. If the diffusivity would be as important as proposed by several investigators, the addition of water and thereby the formation of another phase might hinder the extraction and not force it. The adsorption of the contaminants is very strong and restricts the use of this technique to soil with an moderate content of organic matter. Especially for the transfer of this SFE process in a larger scale the influence of moisture on the extraction process is interesting, because this reduces the required solvent mass and leads to reduced soil treatment costs. Nearly all soils contain water. Therefore, no further equipment or substances are necessary for the use of this entrainer. This has to be taken into account for the design of the soil preparation step. A sophisticated preparation of the soil which leads to an uniform moisture and small particles allows thereby an effective and favorable cleanup of soil contaminated even with low volatile organics. Estimations of the treatment cost show that the extraction with supercritical carbon dioxide can be done with an expense of money in the range of 150 $/ton, depending on the plant scale and the contaminants.

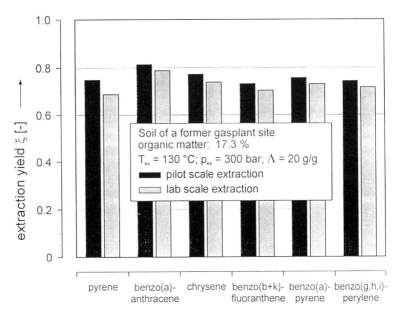

Figure 14. Comparison of the extraction yield in lab and pilot plant scale.

Symbols

c	concentration in fluid phase	T	temperature
K_F	Paramater in Freundlich isotherm	w	content of liquid
m	mass	y	mol fraction
n	Paramater in Freundlich isotherm	ξ	extraction yield
p	pressure	Γ	UV detector signal
q	concentration in soil	Λ	specific solvent ratio
s	entrainer effect		

Literature Cited

(1) Dahmen, N.; Schön, J.; Schmieder, H. *Chemie-Ingenieur-Technik* **1995**, 11, pp 1501.

(2) Laitinen, A.; Michaux, A.; Aaltonen, O. *Environ. Technol.* **1994**, 15, pp 715.

(3) Low, G.K.; Duffy, G.J.; Sharma, M.D.; Chensee, S.; Weir, W.; Tibbet, A.R. *Proc. 3rd Symp. I.S.A.S.F.* **1994** (France), Vol. 3, pp 275.

(4) Roop, R.K.; Akgerman, A. *J. Supercrit. Fluids.* **1989**, 2, pp 51.

(5) Birtigh, A.; Brunner, G. *Proc. 3rd Symp. I.S.A.S.F.* **1994** (France), 1, pp 41.

(6) Debenedetti, P.G. *AIChE Journal* **1990**, 36, pp 1289.

(7) Lütge, C.; Oswald, D.; Schleußinger, A.; Schulz, S. *Terra Tech* **1993**, 3, pp 80.

(8) Lütge, C.; Schleußinger, A.; Schulz, S. *Proc. 3rd Symp. I.S.A.S.F.* **1994** (France), 3, pp. 429.

(9) Schmitt, W.J.; Reid, R.C. *Fluid Phase Equilibria* **1987**, 32, pp 77.

(10) Hawthorne, S.B.; Miller, D.J. *Anal. Chem.* **1987**, 59, 1705-1708.

(11) Schmitt, W.J.; Reid, R.C. *Fluid Phase Equilibria* **1986**, 32 , pp 77.

(12) Burk, R.; Kruus, P. *Can. J. Chem. Eng.* **1992**, 70, pp 403.

(13) Panagiotopoulos, A.Z.; Reid, R.C. In *Equations of State Theories and Applications*; Chao, K.C; Robinson, R.L., Eds.; ACS Symposium Series 300; 1986; pp 571-582.

(14) Panagiotopoulos, A.Z.; Kumar, S.K. *Fluid Phase Equilibria* **1985**, 22, pp 77.

(15) Ghonasghi, D.; Guptha, S.; Dooley, K.M.; Knopf, F.C. *AIChE Journal* **1991**, 6, pp 944.

(16) Macnaughton, S.J.; Foster, N.R. *Ind. Eng. Chem. Res.* **1994**, 33, pp. 2757.

(17) Wiebe, R.; Gaddy, V.L. *J. Am. Chem. Soc.* **1939**, 61, pp 315.

(18) Andrews, A.T.; Ahlert, R.C.; Kosson, D.S. *Env. Prog.* **1990**, 9, pp 204.

(19) Lütge, C.; Reiß, I.; Schleußinger, A.; Schulz, S. *J. Supercrit. Fluids* **1994**, 4, pp. 265.

(20) Kothandaraman, S.; Ahlert, R.C., Venkataramani, E.S.; Andrews, A.T. *Environ. Progress* **1992** , 3, pp 220.

(21) Chiou, C.T. *Environ. Sci. Technol.* **1985**, 19, pp 1196.

(22) Brady, B.O.; Kao, C.-P.C.; Dooley, K.M.; Knopf, F.C. *Ind. Eng. Chem. Res.* **1987**, 2, pp 261.

(23) Dooley, K.M.; Ghonasgi, D.; Knopf, F.C. *Environ. Progress* **1990**, 9, pp 197.

(24) Dzombak, D.A. *Soil Science* **1984**, 5, pp 292.

(25) Hess, R.K.; Erkey, C.; Akgerman, A. *J. Supercrit. Fluids* **1991**, 4, pp 47.

Chapter 14

Estimation of Physicochemical Properties Using Supercritical Fluid Chromatography

Sermin G. Sunol[1], Brad Mierau[2], Irmak Serifoglu[2], and Aydin K. Sunol[2]

[1]Department of Chemical Engineering, Bogazici University, 80815 Bebek, Istanbul, Turkey
[2]Department of Chemical Engineering, University of South Florida, Tampa, FL 33620

The selection of chromatographic techniques in determination of physicochemical properties over conventional static methods is due to quick data turnaround with such systems and readily available commercial equipment. Supercritical fluid chromatography is and can be applied to physicochemical property estimation and its use for this purpose has received considerable attention especially more recently. Moreover, some techniques that are used for determining properties of gas-solid systems using gas chromatography can easily be applied to supercritical fluid chromatography.

Supercritical Fluid Chromatography (SFC) is a rather recent tool that continues to receive considerable attention for chemical analysis, preparative scale separation as well as physicochemical property estimation. Given the existing difficulties in physicochemical property measurement and estimation under supercritical conditions, any SFC based approach deserves a special attention, in part due to its promise in high data turnaround. Although the area is quite new, there is some foundation, from similar GC and HPLC work, to build upon as well as more directly applicable earlier works of Giddings (1) and Kobayashi (2). Excellent reviews are provided by Conder and Young (3) and Laub and Pecsok (4) describing chromatographic methods as well as application of the methods to measurement of physicochemical properties in gas chromatography. Several review papers on physicochemical property measurement by SFC are provided by Schneider and his co-workers (5,6). SFC systems are described by Taylor (7) in detail primarily for analytical purpose and the considerations are valid for physicochemical property estimation as well.

This paper provides an overview of our ongoing research work on physicochemical property determination using SFC as well as a review on the subject. It also includes an analysis of the research performed on the physicochemical property determination on GC and discusses the merits of applying these methods to SFC systems.

Physicochemical Measurements by SFC

Chromatographic experiments at supercritical conditions would provide an effective avenue for determination of several important physicochemical properties. Table I shows a general classification of physicochemical properties which can be measured by SFC.

Table I. Physicochemical Properties which can be Measured by SFC

Equilibrium Properties	Kinetic and Transport Properties	Other Properties

* Fluid Phase Interactions
  - Second Virial Coefficients
* Solution Interactions
  - Partial Molar Volumes
  - Solubilities
* Surface Interactions
  - Adsorption Isotherms
	* Diffusion Coefficients
	* Adsorption and desorption kinetics
	  -Mass Transfer Coefficients
	  -Adsorption and Desorption Rate Constants
		* Molecular masses

	$^\circ$Reaction Rate Constants	

* Properties determined by SFC
$^\circ$ Properties determined by GC and can be determined by SFC

There are three basic types of properties. The first type of properties are equilibrium properties, which can be determined from retention data obtained using the elution technique at infinite dilution. Adsorption isotherms can be determined from experiments at infinite dilution as well as from the concentration - time profile of the migrating peak at finite concentrations. The latter can be determined using the elution technique for reversible adsorption. Also, frontal analysis and combined frontal and elution analysis (elution on a plateau) can be applied for determination of adsorption isotherms. Kinetic and transport properties can be determined from the concentration - time profile of the migrating peak or from the extent to which the peak is broadened as it moves towards column outlet. Properties such as molecular mass can also be determined from infinite dilution experiments.

Capacity Ratio

Definition of Capacity Ratio. The most important property in the theory of chromatographic separation process is the capacity ratio k. Many physicochemical properties can be related to the capacity ratio k. The capacity ratio of the component experimented with is given by:

$$k = \frac{c_s}{c_m} \frac{V_{ts}}{V_{tm}} = K \frac{V_{ts}}{V_{tm}} = \frac{y_s}{y_m} \frac{V_m}{V_s} \frac{V_{ts}}{V_{tm}}$$

where c_s and c_m are concentrations of the solute in the stationary and mobile phases respectively; V_{ts} and V_{tm} are the total volumes of the stationary and mobile phases; K is the partition coefficient; y_s and y_m are mole fractions of the solute in the in the stationary phase and the mobile phases; V_s and V_m are molar volumes of the stationary phase and the mobile phase.

The capacity ratio, k, can be determined relatively easily and accurately from SFC experiments, since in linear chromatography, it is related to the total retention time t of the solute and to the retention time of an inert substance t_0, according to:

$$k = \frac{\left(t - t_0\right)}{t_0}$$

Properties Determined from the Capacity Ratio.

Second Virial Coefficients. Second virial interaction coefficients between the supercritical solvent and the solute can be determined from the slopes of ln k vs P isotherms.

$$\left[\frac{\partial \ln k}{\partial P}\right]_T = \frac{2B - \overline{V}_s^{\infty}}{RT}$$

where B is the second virial interaction coefficient.

Virial coefficients (B) for several systems have been determined by SFC since the mid-1960's. Table II lists second virial coefficients of different solutes in supercritical carbon dioxide up to pressures close to the critical pressure of carbon dioxide. SFC is also promising for the physicochemical study of higher order coefficients (*4*).

Table II. Second Virial Coefficients

Solute	T(°C)	P (atm)	B (-cm³/mol)	Ref.
Acetone	40	1-30	343	8
Benzene	80	1-30	216	9
Carbon tetrachloride	80	1-30	154	8
Chloroform	80	1-30	169	8
Cyclohexane	80	1-30	163	9
Decane	40	20-90	635	8
n-Heptane	80	1-30	177	9
n-Hexane	80	1-30	147	9
Nitroethane	80	1-30	247	8
Nitromethane	80	1-30	188	8
n-Nonane	80	1-30	249	8
n-Octane	80	1-30	277	8
n-Pentane	80	1-30	76	8
Pyridine	80	1-30	207	9
2,2,4-Tri-methylpentane	80	1-30	147	8
Toluene	80	1-30	235	8

Partial Molar Volume in the Stationary Phase. From retention data at infinite dilution of the solute and using a simple thermodynamic model, one can determine partial molar volume of the solute in the stationary phase, provided that the partial molar volume in the mobile phase can be independently determined.

There are two approaches for determining partial molar volume of the solute in the stationary phase, which are shown in Figure 1. There have been attempts for determining partial molar volume of the solute in the stationary phase (10, 11, 12, 13). However, all literature data is collected with coated capillary columns and the procedure neglects pressure dependence of the partial molar volume of the solute in the stationary phase and swelling of the column material. This results in substantial error. Experiments need to be carried out on solid packings or adsorbents in order to be able to determine partial molar volume of the solute in the stationary phase.

Partial Molar Volume in the Mobile Phase. Partial molar volumes in the compressed fluid mixtures are difficult to determine experimentally, especially at very low concentrations, i.e. infinite dilution. There has been attempts to determine partial molar volume of solutes in the mobile phase where partial molar volumes in the stationary phase are calculated (5). On the other hand, if the stationary phase is a

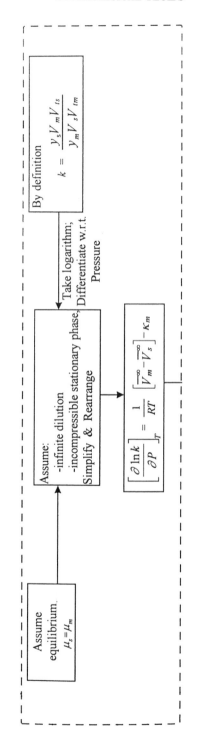

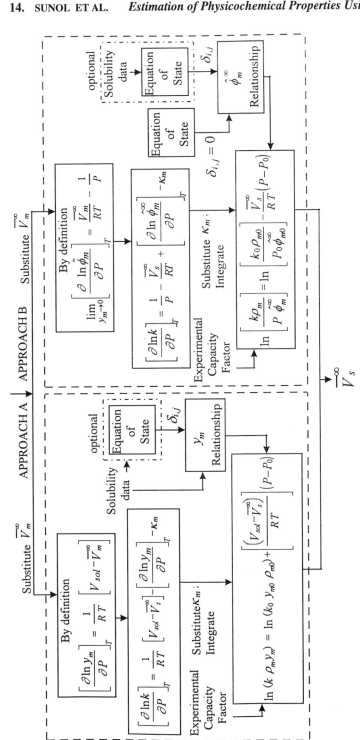

Figure 1. Procedure for calculation of $\overline{V}_s^\infty$.

solid, partial molar volume of the solid is equal to the molar volume of the solid. Provided that the partial molar volume of species in the stationary phase is known, one can calculate the partial molar volume of the solute in the mobile phase using the model and equations given in Figure 1.

Properties Determined Using Various Chromatographic Techniques

Solubilities. Solubility measurements can be performed on a SFC where relatively short fused silica capillary tubing replaces the usual column. Sample is coated into this tubing using the same techniques that are used to coat a stationary phase on an analytical capillary column. Isothermal solubility measurements are made by raising the solvent pressure stepwise from a selected starting point. At each step, the ion current in the mass spectrometer, which is used as the detector, is monitored to determine the quantity of solute in solution (*14, 15*).

Solubility values may also be extrapolated using the last equation in Figure 1, provided that the adsorbent is a solid and partial molar volume of the solute in the stationary phase is equal to the solid molar volume of the solute and the solubility value is known for at least one condition.

Adsorption Isotherms. Another physicochemical investigation, which is important for understanding chromatographic separation as well as any adsorption process is the determination of adsorption isotherms. Research done on SFC for measurement of adsorption isotherms can be divided into two groups according to the two purposes mentioned.

Adsorption Isotherms for SFC Carriers and Capillary Columns. The measurement of the sorption isotherms of the mobile phase components (carrier and modifier) in the stationary phase is important for understanding the retention process in SFC (both analytical and preparative scale). A rather recent review summarizes studies performed towards this purpose (*16*).

Lochmuller and Mink (*17, 18*) studied adsorption isotherms using the peak maxima method and Janssen et al. (*19*) measured sorption isotherms using the breakthrough method. In the peak maxima method, different concentrations of the carrier and solvent mixtures are injected into the column. Capacity ratios are measured for each concentration of the solute and isotherms are calculated from these retention data. In the breakthrough method, a stream of mixed carrier fluid and solute is continuously fed into the column under constant concentration. Then, this mobile phase is instantaneously replaced by a continuous stream of carrier fluid containing solute at different concentrations. The concentration profile is therefore step-shaped.

Studies of supercritical fluid mobile phase component adsorption isotherms using tracer pulse chromatography-mass spectrometry (TCP-MS) have been reported by Parcher and coworkers (*20-24*) and Yonker and Smith (*25, 26*). The systems for which adsorption isotherms of supercritical fluids are determined are given in Table III.

Table III. Adsorption Isotherms Determined by SFC

Adsorbate	Adsorbent	Ref.
Ethyl acetate from supercritical CO_2	Silica	17
Methanol and 1-hexanol from supercritical CO_2	Silica	18
Different modifiers in supercritical CO_2	Octadecyl bonded silica	19
Supercritical pentane in methanol modified supercritical pentane	5% phenylmethylsilica	20
Supercritical CO_2	Silica, octadecyl-silica, cyano-silica, diol-silica	21
Supercritical CO_2	Octadecyl bonded silica	22
Supercritical CO_2	SE-30 polymer, octadecyl bonded silica, silica	23
CO_2 - methanol (binary isotherm)	Octadecyl bonded silica, silica	24
2-propanol from supercritical 2-propanol-CO_2 mixture	Phenyl poly (methylsiloxane) (SE-54)	25
Supercritical CO_2	Phenyl poly (methylsiloxane) (SE-54)	26

In tracer pulse chromatography, the adsorbate flows continuously through the column and a mass isotope of the adsorbate is introduced into the column at a trace amount. (For example oxygen labeled carbon dioxide ($C^{18}O_2$)). The tracer isotope is detected at the effluent of the column using a mass spectrometer. Also, retention of an unretained (inert) component is detected. The amount of the adsorbate (N_i^{sp}) in the stationary phase is calculated from the equation:

$$N_i^{sp} = \frac{V_n y_i P}{ZRT}$$

where V_n the net retention volume, y_i is the mole fraction of the solvent molecule under study, P is the system pressure, Z is the compressibility of the fluid (determined from an appropriate equation of state), R is the gas constant and T is the system temperature.

From the net retention time of the labeled isotope and the inert component, the net retention volume for the isotope can be calculated:

$$V_n = \left[(t-t_0)/t_0 \right] \pi r^2 L$$

where r is the column radius, L is the column length, t is the total retention time of the solute and t_0 is the retention time of an inert substance.

The aforementioned studies are all limited to adsorption of single component with the one exception of a binary isotherm in reference 24. Studies on the determination of adsorption isotherms could be extended to multi-component (even reactive) systems.

Tracer pulse chromatography is preferred because of the ease of calculations, although it has disadvantages of demanding a highly expensive detection system and complications as a result of having an isotope in the system. An alternative to tracer pulse chromatography employed in GC is concentration perturbation chromatography. Its application to SC systems would be more difficult due to the complexity of PVT behavior for high pressure systems.

Adsorption Isotherms for Supercritical systems other than SFC. The isotherms for different supercritical systems of interest can also be determined using different chromatographic techniques which are widely applied to GC. Akgerman and coworkers (27) determined adsorption isotherms of naphthalene, phenanthrene, hexachlorobenzene, and pentachlorophenol on activated carbon in the presence of carbon dioxide using an experimental system based on frontal analysis chromatography. Such studies can also be performed on a SFC where the adsorbent is packed into the chromatographic column. Any chromatographic method applied to GC for isotherm determination such as peak maxima method, peak profile method, step profile method, minor disturbance method (non-equilibrium methods) as well as equilibrium methods, can be applied to SFC in order to determine sorption isotherms at supercritical conditions.

Diffusion Coefficients. Diffusivity determinations with open tubes, strictly speaking, not chromatographic in nature since no adsorption/desorption or partitioning in two phases are involved, provide best diffusion coefficient data available. The experimental system used is a chromatograph, and diffusion occurs in an empty, inert column on which the fluid phase is not supposed to be adsorbed. The fluid in which the solute diffuses flows continuously through the empty column and the solute is introduced into the column at one end, at the other end the effluent concentration is detected.

The theory of diffusion in flowing fluids is first given by Taylor and subsequently formalized by Aris. According to Aris, a sharp band of solute, which is allowed to dissolve in a solvent flowing laminarly in an empty tube, can be described in the limit of a long column as a Gaussian distribution, the variance of which, σ, in length units is

$$\sigma(x)^2 = 2\, D_{eff}\; t$$

where t is the time of migration of the peak, D_{eff} is an effective diffusion coefficient, given by

$$D_{eff} = D_{12} + \frac{r^2 \, u^2}{48 \, D_{12}}$$

where u is the average solvent velocity, r is the inner radius of the tube, and D_{12} is the binary diffusion coefficient. The first term describes the longitudinal diffusion in the axial direction. The second term is called the Taylor diffusion coefficient and describes band broadening due to the parabolic flow profile and therefore radial diffusion. The height equivalent to a theoretical plate, H, is a measure of the relative peak broadening and is defined as

$$H = \sigma(x)^2 / L$$

where L is the length of the tubing. Combining the equations above yields

$$H = \frac{2D_{12}}{u} + \frac{r^2 u}{24 D_{12}}$$

and bulk diffusivity can be calculated from the above equation.

Another more accurate way of calculating the diffusivity from broadened peak is to fit experimental concentration versus time data to the model given by Taylor and Aris in order to calculate binary diffusivities (*28, 29, 30*).

Supercritical systems for which bulk diffusivities are determined using the peak broadening method are given in Table IV (*28-53*).

Adsorption Kinetics. The use of SFC to examine adsorption equilibria has been described. The advantages of the chromatographic technique are also relevant for kinetic studies. The kinetics of fluid-solid adsorption process are reflected in the peak shape. The width of the peak is governed both by dispersion and by the adsorption desorption kinetics. In order to separate the broadening due to dispersion from adsorption, two approaches are used: Either experimental conditions are adjusted such that dispersion is negligible and peak broadening is attributed to adsorption only, or additional experiments are conducted with variable flow rate and particle size to determine contributions of both dispersion and adsorption. In the first approach, the kinetic parameter estimated is time spent on the adsorption site (*53, 54*) and in the second approach, the rate constant for adsorption and desorption can be determined (*55-57*).

Lee and Holder (*58*) used the latter approach for the determination of mass transfer coefficients in supercritical fluids. They describe the determination of mass transfer coefficients in supercritical fluids by regression of these parameters as model coefficients from experimental data. By assuming a linear adsorption isotherm, one can obtain initial estimates of the chromatographic model parameters by use of the moment method which results in

Table IV. Binary Diffusivities in Supercritical Fluid

Solvent	Solute	T (° C)	P (atm)	D_{12} *10^5 (cm^2/s)	Ref
2,3-dimethylbutane	benzene	250-275	53-157	25.0-40.4	40
2,3-dimethylbutane	naphthalene	250-275	53-157	19.9-32.1	40
2,3-dimethylbutane	phenanthrene	250-275	53-157	17.2-26.7	40
2,3-dimethylbutane	toluene	250-275	53-157	23.0-38.7	40
2-propanol	benzene	235-263	72-98	21.6-34.4	39
2-propanol	naphthalene	235-263	72-98	17.5-26.1	39
2-propanol	n-decane	235-263	72-98	15.3-23.8	39
2-propanol	n-tetradecane	235-263	72-98	12.3-18.7	39
2-propanol	phenanthrene	235-263	72-98	14.6-22.4	39
2-propanol	toluene	235-263	72-98	19.7-31.3	39
CClF$_3$	1,3-dibromobenzene	10-65	30-148	7.0-18.8	32
CClF$_3$	1,4-dimethylbenzene	10-65	30-148	10.5-24.2	32
CClF$_3$	2-propanone	10-65	30-148	14.5-40.2	32
CO$_2$	α-tocopherol	35-45	100-248	4.5-5.7	31
CO$_2$	1,3,5-trimethyl benzene	40	79-158	10.0-17.9	51
CO$_2$	1,3,5-trimethylbenzene	30	113-187	8.7-10.6	47
CO$_2$	1,3,5-trimethylbenzene	40	158-247	9.1-10.5	47
CO$_2$	1,3,5-trimethylbenzene	60	128-261	11.6-20.9	47
CO$_2$	2,2,4,4-tetramethyl-3-pentanone	40	99-158	14.3-22.2	49
CO$_2$	2,4-dimethyl-3-pentanone	40	99-158	15.8-24.5	49
CO$_2$	2-butanone	41	94-178	17.5-34.6	49
CO$_2$	2-heptanone	41	94-178	14.0-29.2	49
CO$_2$	2-methyl-1,4 naphthalendione (vitamin K3)	40	158	9.3	41
CO$_2$	2-nonanone	41	94-178	13.1-21.6	49
CO$_2$	2-pentanone	41	94-178	15.3-33.0	49
CO$_2$	2-propanone	40	99-158	18.4-29.2	49
CO$_2$	3-heptanone	41	99-178	2.3-2.8	48
CO$_2$	3-pentanone	41	99-178	2.4-2.9	48
CO$_2$	3-pentanone	40	99-158	17.2-26.8	49
CO$_2$	3-pentanone	41	94-178	16.8-31.7	49
CO$_2$	4-cyano-4-n-pentoxybiphenyl	41	128-178	8.6-10.6	49
CO$_2$	4-heptanone	40	99-158	15.9-24.3	49
CO$_2$	5-nonanone	41	99-178	2.2-2.7	48
CO$_2$	5-nonanone	41	94-178	12.0-22.3	49
CO$_2$	6-undecanone	41	94-178	11.2-19.7	49
CO$_2$	acetone	30	113-187	11.7-14.0	35
CO$_2$	acetone	40	158-246	12.7-14.7	35
CO$_2$	acetone	60	128-261	15.2-27.4	35
CO$_2$	adamantanone	40	99-158	14.1-14.4	49
CO$_2$	benzene	40	148-296	11.7-15.2	33
CO$_2$	benzene	40	158	12.4	30
CO$_2$	benzene	40	247	11.5	30
CO$_2$	benzene	36	90-247	0.95-1.8	43
CO$_2$	benzene	50	148-345	10.8-15.6	44
CO$_2$	benzene	60	148-345	11.6-18.2	44
CO$_2$	benzene	40	88-134	13.0-17.5	46
CO$_2$	benzene	50	88-134	14.2-25.5	46

Table IV. *Continued*

Solvent	Solute	T (° C)	P (atm)	$D_{12}*10^5$ (cm²/s)	Ref
CO_2	benzene	30	113-187	11.5-13.8	47
CO_2	benzene	40	109-246	12.8-21.3	47
CO_2	benzene	60	128-261	14.8-25.8	47
CO_2	benzene	40	79-158	15.8-31.6	50
CO_2	benzene	40	79-158	12.9-29.9	51
CO_2	benzene	40	148-345	9.0-13.0	44
CO_2	benzene	40	160	12.8	28
CO_2	benzoic acid	35-45	100-248	10.9-14.3	31
CO_2	biphenyl	100	100	0.37	45
CO_2	biphenyl	125	100	0.50	45
CO_2	C10:00 ethyl ester	35-45	95-208	7.1-14.3	35
CO_2	C14:00 ethyl ester	35-45	95-208	6.1-12.0	35
CO_2	C15 n-alkane	100	100	0.34	45
CO_2	C15 n-alkane	125	100	0.40	45
CO_2	C16 n-alkane	100	100	0.33	45
CO_2	C16 n-alkane	125	100	0.39	45
CO_2	C16-C24 unsaturated fatty acidmethylester	40	158	5.6-6.8	29
CO_2	C17 n-alkane	125	100	0.37	46
CO_2	C18 n-alkane	100	100	0.30	45
CO_2	C18 n-alkane	125	100	0.37	45
CO_2	C18:00 ethyl ester	35-45	95-208	5.5-10.5	34
CO_2	C20:5 methyl ester	35-45	95-208	5.5-10.6	34
CO_2	C22:00 ethyl ester	35-45	95-208	5.2-10.1	31
CO_2	C22:6 ethyl ester	35-45	95-208	5.4-10.2	34
CO_2	C22:6 methyl ester	35-45	95-208	5.4-10.3	34
CO_2	C4:00 ethyl ester	35-45	95-208	9.4-18.5	35
CO_2	C8:00 ethyl ester	35-45	95-208	7.6-15.0	35
CO_2	caffeine	40	79-158	7.1-7.9	50
CO_2	carbontetrachloride	36	104-143	0.86-0.94	43
CO_2	chrysene	30	188	7.1	47
CO_2	chrysene	40	158-247	7.4-8.6	47
CO_2	chrysene	60	263	9.6	47
CO_2	cis-11-eicosenoic acid methyl ester	40	158	6.2	30
CO_2	cis-11-eicosenoic acid methyl ester	40	247	5.2	30
CO_2	cis-jasmone	40	158	8.8	30
CO_2	cis-jasmone	40	247	7.5	30
CO_2	citronellol	60	148-296	8.1-8.8	33
CO_2	cycloheptanone	41	99-178	2.3-2.8	48
CO_2	cyclononanone	41	99-178	2.2-2.7	48
CO_2	cyclopentanone	41	99-178	2.4-2.9	48
CO_2	dl-limonene	40	158-247	8.7-10.9	30
CO_2	dodecanone	125	100	0.42	45
CO_2	dodecanone	100	100	0.39	45
CO_2	erucic acid methyl ester	40	158	5.9	30
CO_2	erucic acid methyl ester	40	247	5.1	30
CO_2	ethylbenzene	40	148-345	8.8-12.1	44
CO_2	ethylbenzene	50	148-345	9.8-14.7	44
CO_2	ethylbenzene	60	148-345	10.8-17.2	44
CO_2	geraniol	60	148-296	6.9-8.4	33

continued on next page

Table IV. *Continued*

Solvent	Solute	T (° C)	P (atm)	D_{12} *10^5 (cm²/s)	Ref
CO_2	geraniol	60	148-296	7.5-8.5	33
CO_2	glycerol triolate	35-45	100-248	3.5-4.6	31
CO_2	indole (2,3-benzopyrrole)	40	158	10.4	30
CO_2	indole (2,3-benzopyrrole)	40	246.7	9.2	30
CO_2	isopropylbenzene	40	148-345	1.6-8.1	44
CO_2	isopropylbenzene	50	148-345	3.4-9.4	44
CO_2	isopropylbenzene	60	148-345	5.9-9.9	44
CO_2	linalool	60	148-296	8.0-8.7	33
CO_2	linoleic acid	41	178-128	7.8-10.0	49
CO_2	linoleic acid methyl ester	35-55	187	6.5	28
CO_2	linoleic acid methyl ester	35-45	138-332	5.4-6.3	29
CO_2	m-cresol	40	88-134	10.0-13.6	46
CO_2	m-cresol	50	88-134	12.0-14.5	46
CO_2	methylmistrate	125	100	0.37	45
CO_2	methylmistrate	100	100	0.28	45
CO_2	myristoleic acid methyl ester	40	158	7.2	30
CO_2	myristoleic acid methyl ester	40	247	6.0	30
CO_2	n-4-methoxybenzy lidene-4-n-butylaniline	40	158-118	10.0-12.3	49
CO_2	naphthalene	40	158	11.2	30
CO_2	naphthalene	40	247	9.4	30
CO_2	naphthalene	30	113-187	9.3-11.3	47
CO_2	naphthalene	40	158-246	9.6-15.6	47
CO_2	naphthalene	60	128-261	12.3-21.9	47
CO_2	naphthalene	40	79-158	11.2-24.5	50
CO_2	naphthalene	40	160	11.6	28
CO_2	nerol	60	148-296	7.6-7.7	33
CO_2	n-propylbenzene	40	148-345	1.6-8.7	44
CO_2	n-propylbenzene	50	148-345	3.5-9.7	44
CO_2	n-propylbenzene	60	148-345	5.9-10.0	44
CO_2	n-propylbenzene	40	79-158	10.5-19.9	51
CO_2	oleic Acid	35-45	100-248	4.8-5.6	31
CO_2	oleic acid	40	158-118	10.8-15.4	49
CO_2	pentadecanone	100	100	0.29	45
CO_2	pentadecanone	125	100	0.35	45
CO_2	phenanthrene	30	113-187	8.5-8.9	47
CO_2	phenanthrene	40	158-247	8.2-9.6	47
CO_2	phenanthrene	60	128-261	10.7-18.4	47
CO_2	phenol	40	79-158	12.6-25.1	50
CO_2	phenylacetic acid	35-40	95-185	8.6-15.2	36
CO_2	pyrene	30	115	9.3	47
CO_2	pyrene	40	247	7.9	47
CO_2	pyrene	60	163	14.4	47
CO_2	squalene	41	128-178	6.8-7.8	49
CO_2	stearic acid	40	118-158	11.0-12.8	49
CO_2	terpineol	60	148-296	7.5-10.2	33
CO_2	terpineol	60	148-296	9.0-10.0	33
CO_2	toluene	40	148-296	10.1-13.1	33
CO_2	toluene	40	148-345	9.3-12.2	44
CO_2	toluene	50	148-345	10.3-15.3	44

Table IV. *Continued*

Solvent	Solute	T (° C)	P (atm)	$D_{12}*10^5$ (cm²/s)	Ref
CO_2	toluene	45	130-300	11.9-13.0	52
CO_2	toluene	60	148-345	11.4-17.2	44
CO_2	vanilin	35-40	95-185	8.4-14.2	36
CO_2	vitamin A acetate	40	158	5.9	30
CO_2	vitamin A acetate	40	247	5.5	30
CO_2	vitamin E	40	158	5.5	30
CO_2	vitamin K1	40	158	5.4	30
CO_2	vitamin K1	40	247	4.9	30
CO_2	vitamin K3	40	158	9.3	30
CO_2	vitamin K3	40	247	7.9	30
$CO_2/1\%CH_3CN$	benzene	40	88-134	12.0-14.4	46
$CO_2/1\%CH_3CN$	benzene	50	88-134	13.9-19.9	46
$CO_2/1\%CH_3CN$	m-cresol	40	88-134	8.9-13.3	46
$CO_2/1\%CH_3CN$	m-cresol	50	88-134	9.7-14.0	46
$CO_2/1\%C_4H_6O_3$	m-cresol	40	88-134	9.8-13.3	46
$CO_2/1\%C_4H_6O_3$	m-cresol	50	88-134	11.6-13.5	46
$CO_2/5\%CH_3CN$	benzene	40	88-134	9.0-13.0	46
$CO_2/5\%CH_3CN$	benzene	50	88-134	11.0-15.0	46
$CO_2/5\%CH_3CN$	m-cresol	40	88-134	9.5-13.0	46
$CO_2/5\%CH_3CN$	m-cresol	50	88-134	10.5-14.0	46
EtOH	benzene	243-281	61-95	22.1-36.6	37
EtOH	mesitylene	243-281	61-95	16.7-28.5	37
EtOH	naphthalene	243-281	61-95	18.2-29.6	37
EtOH	phenanthrene	243-281	61-95	15.7-25.3	37
EtOH	toluene	243-281	61-95	21.1-33.3	37
n-hexane	benzene	249-270	155-306	29.3-42.5	38
n-hexane	mesitylene	249-270	155-306	23.1-33.0	38
n-hexane	naphtalene	249-270	155-306	24.1-33.8	38
n-hexane	phenanthrene	249-270	155-306	20.7-28.4	38
n-hexane	p-xylene	249-270	155-306	25.5-36.0	38
n-hexane	toluene	249-270	155-306	27.3-38.6	38
Propane	benzene	110	88-134	14.7-21.5	46
Propane	benzene	130	88-134	20.0-23.0	46
Propane	m-cresol	110	88-134	14.7-15.5	46
Propane	m-cresol	130	88-134	15.6-19.0	46
Propane/H_2O	benzene	110	88-134	17.0-20.3	46
Propane/H_2O	benzene	130	88-134	19.5-24.2	46
Propane/H_2O	m-cresol	110	88-134	14.7-16.5	46
Propane/H_2O	m-cresol	130	88-134	16.8-19.3	46
SF_6	1,3,5-trimethylbenzene	10-65	30-148	7.4-30.7	32
SF6	1,4-dimethylbenzene	10-65	30-148	4.5-42.3	32
SF_6	benzene	10-65	30-148	9.5-56.0	32
SF_6	linoleic acid methyl ester	55	142-197	4.0-4.5	42
SF_6	methylbenzene	10-65	30-148	9.0-45.0	32
SF_6	tetrachloromethane	10-65	30-148	7.8-41.0	32
SF_6	toluene	55	142-197	8.1-9.0	42
SF_6	vitamin K3	55	142-197	6.2-6.9	42

$$\mu_1 = \frac{L}{u}\left(1+\delta_0\right) + \frac{\tau_0}{2}$$

$$\mu_2 = \frac{2L}{u}\left[\frac{E_A}{\varepsilon_p}\left(1+\delta_0\right)^2\frac{1}{u_2}+\delta_1\right] + \frac{\tau_0^2}{12}$$

where

$$\delta_0 = \frac{\varepsilon_p}{m}\left(1+\frac{\rho_p K_A}{\varepsilon_p}\right)$$

$$\delta_1 = \frac{\varepsilon_p}{m}\left[\frac{R^2\varepsilon_p}{15}\left(1+\frac{\rho_p K_A}{\varepsilon_p}\right)^2\left(\frac{1}{D_e}+\frac{5}{k_f R}\right)\right] + \frac{\rho_p K_A^2}{\varepsilon_p k_{ads}}$$

with the experimental values for the first and second moments obtained from the response curves;

$$\mu_1 = \frac{\int_0^\infty \tau c\,(L,\,\tau)\,d\,\tau}{\int_0^\infty c\,(L,\,\tau)\,d\,\tau}$$

$$\mu_2 = \frac{\int_0^\infty \left(\tau - \mu_1\right)^2 c\,(L,\,\tau)\,d\,\tau}{\int_0^\infty c\,(L,\,\tau)\,d\,\tau}$$

The equilibrium constant (K_a), the axial dispersion coefficient (E_a), the mass transfer coefficient (k_f), and the adsorption rate constant (k_{ads}) are determined from the first and second moments described by the preceding equations. The authors employed the resulting values as initial guesses in the regression of breakthrough data against a model based on the linear driving force concept and Heaviside step function.

Techniques that are Applied to Determine Properties of GC Systems and are Applicable to SFC

As listed in Table I, chemical reaction kinetics has been widely determined using GC, however, so far there are no application of the methods to SFC.

GC is used in order to determine reaction kinetics with special emphasis on catalytic reactions.   Since catalytic reactions under supercritical conditions are important, GC techniques can be applied to SFC.   SFC can be used in four different types of kinetic study.   In the first one, a precolumn reactor may be used downstream from the injection system and upstream from the column.   Reaction products can be identified easily and reaction kinetics can be followed by varying the carrier gas flow.   The SFE/SFC system conventionally used for analysis can be employed for the aforementioned study.

The second method employed for the study of reaction kinetics is "sample vacancy chromatography".   It is based on a precolumn reactor rather than an on-column reactor.   Nevertheless, it does not require a second analytical column.   In this technique, the sample is mixed with a fixed proportion of carrier fluid and passed through the chromatographic column.   After steady state is reached, pure carrier is introduced into the column so that negative peaks for each component are obtained.

In the other two methods, the column becomes the so called "on-column reactor" and combines the function of reactor and separator.   In the "stopped flow chromatography", the fluid flow through the column is stopped for certain time intervals during elution, during which the reaction proceeds and chromatographic process is stopped.   Information about reaction kinetics can be extracted from peak areas of the products corresponding to stopped flow time intervals.   In the last method, reaction kinetics can be determined from distorted shape or peak areas of the reaction chromatogram.

List of Symbols

a_i, a_j	pure component parameter
a_{mix}	mixture parameter
B	second virial interaction coefficient
c_m	concentration of the solute in the mobile phase
c_s	concentration of the solute in the stationary phase
D_{eff}	effective diffusion coefficient
D_{12}	binary diffusion coefficient
E_a	axial dispersion coefficient
H	height equivalent to a theoretical plate
k	capacity ratio
k_{ads}	adsorption rate constant
k_f	mass transfer coefficient
K	partition coefficient
K_a	equilibrium constant
L	column length
m	ratio of bed void fraction, $\varepsilon/(1-\varepsilon)$
N_i^{sp}	amount of the adsorbate in the stationary phase
P	pressure

P_{sub}	sublimation pressure of the solute
r	column radius
R	universal gas constant
t	retention time of the solute
t_0	retention time of the nonadsorbed (inert) compound
T	temperature
u	interstitial velocity
V_{sol}	molar volume of the solid solute
V_m	molar volume of the mobile phase
V_n	net retention volume
$\overline{V}_m^\infty$	partial molar volume of solute in the mobile phase at infinite dilution
V_s	molar volume of the stationary phase
$\overline{V}_s^\infty$	partial molar volume of solute in the stationary phase at infinite dilution
V_{tm}	total volume of the mobile phase
V_{ts}	total volume of the stationary phase
V_m	infinite-dilution partial molar volume of the solute in the mobile phase
V_s	infinite-dilution partial molar volume of the solute in the stationary phase
y_i	mole fraction of the solvent molecule
y_m	mole fraction of the solute in the mobile phase
y_s	mole fraction of the solute in the stationary phase
Z	compressibility factor

Greek Symbols:

δ_{ij}	binary interaction parameters
ε_p	particle void fraction
$\hat{\phi}_m^\infty$	infinite dilution fugacity coefficient of the solute in the mobile phase
κ_m	isothermal compressibility of the mobile phase
μ_m	chemical potential of the mobile phase
μ_s	chemical potential of the stationary phase
μ_1	first absolute moment
μ_2	second central moment
ρ_m	molar density of the mobile phase
ρ_p	particle density
σ	peak variance
τ_0	time of duration of the injection of absorbable gas
τ	time

Literature Cited

1. Giddings, J. C.; Mallik, K. L. *Ind. Eng. Chem.*, **1967**, *59*, 18.
2. Kobayashi, R.; Kragas, T. *J. Chromatogr. Sci.*, **1985**, *23*, 11.
3. Conder, J. R.; Young, C. L. *Physicochemical Measurements by Gas Chromatography;* John Wiley & Sons: Chichester, 1979.
4. Laub, R. J.; Pecsoc, R. L. *Physicochemical Applications of Gas Chromatography;* John Wiley & Sons: New York, 1978.
5. Van Wasen, U.; Swaid, I.; Schneider, G. M. *Angew. Chem. Int. Ed. Engl.* **1980**, *19*, 575.
6. Schneider, G. M. In *Analysis with Supercritical Fluids*; Wenklawiak, B., Ed.; Springer Verlag: Heidelberg, 1992, pp 9-31.
7. Taylor, L. T.; Schweighardt, F. K. presented at ACS Symposium on Supercritical Fluids, New Orleans, March 1996.
8. Vigdergauz, M. S.; Semkin, V. I. *Russian Chemical Reviews*, **1971**, *40*, 533.
9. Vigdergauz, M. S.; Semkin, V. I. *J. Chromatogr.,***1971**, *58*, 95.
10. Yonker, C. R.; Smith, R. D. *J. Phys. Chem.* **1988**, *92*, 1664.
11. Yonker, C. R.; Smith, R. D. *J. Chromatogr.*, **1988**, *459*, 183.
12. Gonenc, Z. S.; Akman, U.; Sunol, A. K. *Can. J. Chem. Eng.*, **1995**, *73*, 267.
13. Gonenc, Z. S.; Akman, U.; Sunol, A. K. *J. Chem. Eng. Data*, **1995**, *40*, 799.
14. Bruno, T. J. In *Supercritical Fluid Technology: Reviews in Modern Theory and Applications;* Bruno, T. J.; Ely, J. F., Eds.; CRC Press: Boca Raton, 1991, 293-323.
15. Smith, R. D.; Udseth, H. R.; Wright, R. W. *Sep. Sci. Technol.*, **1987**, *22*, 1065.
16. Yonker, C. R.; Smith, R. D. *J. Chromatogr.*, **1991**, *550*, 775.
17. Lochmuller, C. H.; Mink, L. P. *J. Chromatogr.*, **1987**, *409*, 55.
18. Lochmuller, C. H.; Mink, L. P. *J. Chromatogr.*, **1989**, *471*,357.
19. Janssen, J. G. M.; Schoenmakers, P. J.; Cramers, C. A. *J. High Resolut. Chromatogr.*, **1989**, *12*, 645.
20. Strubinger, J. R.; Selim, M. I. *J. Chromatogr. Sci.*, **1988**, *26*, 579.
21. Parcher, J. F.; Strubinger, J. R. *J. Chromatogr.*, **1989**, *479*, 251.
22. Strubinger, J. R.; Parcher, J. F. *Anal. Chem.*, **1989**, *61*, 951.
23. Strubinger, J. R.; Song, H.; Parcher, J. F. *Anal. Chem.*, **1991**, *63*, 98.
24. Strubinger, J. R.; Song, H.; Parcher, J. F. *Anal. Chem.*, **1991**, *63*,104.
25. Yonker, C. R.; Smith, R. D. *Anal. Chem.*, **1989**, *61*, 1348.
26. Yonker, C. R.; Smith, R. D. *J. Chromatogr.* **1990**, *505*, 139.
27. Madras, G.; Erkey, C.; Akgerman, A. *Ind. Eng. Chem. Res.*, **1993**, *32*, 1163.
28. Funazukuri, T.; Hachitsu, S.; Wakao, N. *Anal. Chem.*, **1989**, *61*, 118.
29. Funazukuri, T.; Hachitsu, S.; Wakao, N. *Ind. Eng. Chem. Res.*, **1991**, *30*, 1323.
30. Funazukuri, T.; Ishiwata, Y.; Wakao, N. *AIChE Journal*, **1992**, *38*, 1761.
31. Catchpole, O. J.; King, M. B. *Ind. Eng. Chem. Res.*, **1994**, *33*,1828
32. Kopner, A.; Hamm, A.; Ellert, J.; Feist, R.; Schneider, G. M. *Chem. Eng. Sci.*, **1987**, *42*, 2213.

33. Spicka, B.; Alessi, P.; Cortesi, A.; Kikic, I.; Macnaughton, S. In *Proceedings of the 3rd International Symposium on Supercritical Fluids;* Brunner, G.; Perrut, M., Eds.; I.N.P.L.A.R.: Nancy Cedex, 1994, Vol. 1; pp 301-306.
34. Liong, K. K.; Wells P. A.; Foster, N. R. *Ind. Eng. Chem. Res.*,**1991**, *30*, 1329.
35. Liong, K. K.; Wells, A. P.; Foster, N. R. *Ind. Eng. Chem. Res.*,**1992**, *31*, 390.
36. Wells, T.; Foster, N. R.; Chaplin, R. P. *Ind. Eng. Chem. Res.*,**1992**, *31*, 927.
37. Sun, C. K. J.; Chen, S. H. *AIChE J*, **1986**, *32*, 1367.
38. Sun, C. K. J.; Chen, S. H. *Chem. Eng. Sci.*, **1985**, *40*, 2217.
39. Sun, C. K. J.; Chen, S. H. *Ind. Eng. Chem. Res.*, **1987**, *26*, 815.
40. Sun, C. K. J.; Chen, S. H. *AIChE J.*, **1985**, *31*, 1904.
41. Funazukuri, T.; Ishiwata, Y.; Wakao N. *J. Chem. Eng. Japan*, **1991**, *24*, 387.
42. Funazukuri, T.; Nishimoto, N.; Wakao N. In *Proceedings of the 3rd International Symposium on Supercritical Fluids;* Brunner, G.; Perrut, M., Eds.; I.N.P.L.A.R.: Nancy Cedex, 1994, Vol. 1; pp 29-33.
43. Erkey, C.; Gadalla, H.; Akgerman, A. *J. Supercrit. Fluids*, **1990**, *3*, 180.
44. Suarez, J. J.; Bueno, J. L.; Medina, I. *Chem. Eng. Sci.*, **1993**, *48*, 2419.
45. Janak, K.; Hagglund, I.; Blomberg, L. G.; Bemgard, A. K.; Colmsjo, A. L .*J. Chrom.*, **1992**, *625*, 311.
46. Olesik, S. V.; Woodruff, J. L. *Anal. Chem.*, **1991**, *63*, 670.
47. Sassiat, P. R.; Mourier, P.; Caude, M. H.; Rosset, R. H. *Anal. Chem.*, **1987**, *59*,
48. Dahmen, N.; Dulberg, A.; Schneider, G. M. *Ber. Bunsenges. Phys. Chem,*, **1990**, *94*, 384.
49. Dahmen, N.; Kordiowski, A.; Schneider, G. M. *J. Chromatogr.*, **1990**, *505*, 169.
50. Feist, R.; Schneider, G. M. *Sep. Sci. Tech.*, **1982**, *17*, 261.
51. Swaid, I.; Schneider, G. M. *Ber. Bunsenges. Phys. Chem.*, **1979**, *83*, 969.
52. Bruno, T. J. *J. Resarch Natl. Inst. Stand. Tech.*, **1989**, *94*, 105.
53. Eberly, P. E.; Spencer, H. *Trans. Faraday Soc.*, **1961**, *57*, 289.
54. Eberly, P. E. *J. Appl. Chem.*, **1964**, *14*, 330.
55. Schneider, P.; Smith, J. M. *AIChE. J.*, **1968**, *14*, 762.
56. Padberg, G.; Smith, J. M. *J. Catal.*, **1968**, *12*, 172.
57. Adrian, J, C.; Smith, J. M., *J. Catal.*, **1970**, *18*, 57.
58. Lee, C. H.; Holder, G. D. *Ind. Eng. Chem. Res.*, **1995**, *34*, 906.

ENVIRONMENTAL APPLICATIONS

Chapter 15

Supercritical Fluids in Environmental Remediation and Pollution Prevention

A. Akgerman

Department of Chemical Engineering, Texas A&M University,
College Station, TX 77843–3122

During the last decade use of supercritical fluids in environmental applications have increased due to their unique properties. Technologies have already been developed for extraction of organic compounds from aqueous and solid environmental matrices and research on extraction of metals is in progress. In most applications, supercritical carbon dioxide is the solvent of choice because it is environmentally benign, safe, and abundant at a low cost. Recent research focus is on use of supercritical fluids as separation and reaction media. Especially as the reaction media, supercritical fluids offer properties that may significantly affect reaction rates and selectivities. The most important property is the control of solvency power through density. This aspect may be used to eliminate side reactions, exceed thermodynamic yield limitations, or control polymer chain lengths and molecular weight distributions. In the separations area, supercritical fluids can be exploited for adsorptive separations of structurally very similar compounds and for selective extraction of thermally labile compounds from natural products.

A supercritical fluid (SCF) is a fluid at conditions above its critical temperature and pressure. Interest in the extraction of solid and liquid media by SCFs have increased during the last decade due to: (1) environmental problems associated with common solvents (mostly chlorinated hydrocarbons); (2) the increasing cost of energy intensive separation processes; and (3) the inability of conventional separation processes to provide the necessary separations needed in emerging new industries. The attractive physicochemical properties of SCFs qualify them as a viable alternative to conventional solvents used in extraction processes.

At temperatures and pressures above its critical point a pure substance exists in a state that exhibits gas-like and liquid-like properties. The fluid's density would be very close to that of a liquid, the surface tension is very close to zero, the diffusivity and viscosity have a value somewhat in between that of a liquid and a gas, and most important the solvent power of a SCF can be varied over a very wide range by small variations in temperature and/or pressure in the supercritical region. These properties result in several advantages in extraction such as ease of solvent recovery, elimination of residual solvent in the extracted medium, lower pressure drops, and higher mass transfer rates.

In extraction of a solute from a matrix (such as water) the choice of the solvent depends on two criteria, its immiscibility with the matrix and the solubility of the solute in the solvent. The solubility in SCFs is a strong function of density. In the vicinity of the critical point, $1 < T_r < 1.1$ and $1 < P_r < 2$, the density is a very strong function of both the temperature and the pressure. The solvent characteristics of a SCF can therefore be adjusted as desired, an important advantage compared to conventional solvents. At extraction conditions the solvent power would be high so the SCF fluid can remove the solute from the matrix and at the separation/solvent recovery stage (where the solute is removed from the solvent) the solvent power would be reduced to close to zero.

The use of a compressed gas in a separation process was first proposed for oil deasphalting (*1*). Although the process was not strictly supercritical, it did take advantage of the change in solubility associated with a pressure reduction. Later, Elgin and Weinstock (*2*) reported on a phase-splitting technique for recovery of methyl-ethyl-ketone (MEK) from water using supercritical ethylene. Since then SCFs have been used as solvents to extract a variety of solid and liquid matrices such as coal (*3*), caffeine from coffee (*4, 5*), tobacco (*6*), fruit aromas (*7*), and alcohols from water (*8, 9*). Various symposia proceedings and books on supercritical extraction phenomena are available in the literature. The broad range of organic solutes that can be extracted from aqueous and solid waste (wastewater, contaminated soils, sludges etc.) by SCFs as well as the availability of inexpensive, readily available, and non-toxic SCF solvents such as CO_2, has directed attention to this process as a viable method for removing toxic organic compounds.

Supercritical fluid technology is also rapidly finding application as an emerging technology in pollution prevention and environmentally friendly manufacturing. The supercritical fluid technology has the potential to provide two very important advantages: (a) replacement of toxic organic solvents used in chemical processes by non-toxic and environmentally acceptable supercritical fluids, and (b) to yield better selectivities to the desired compounds or more pure products, thereby eliminating costly secondary separations and minimizing waste product accumulation. Some applications are chemical reactions that require organic solvents, chemical separations that are accomplished by dissolution in an organic solvent, and chemical extractions that use organic solvents.

In the following discussion aqueous waste and solid waste are treated under separate headings since the fundamentals of the technology as well as the application techniques are slightly different. This is followed by a section on applications in pollution prevention area. Significant amount of work is done on supercritical water oxidation and recent reviews are available, therefore the discussion of this subject is not included in this manuscript.

Supercritical Extraction of Aqueous Environmental Media

The advantages of employing SCF extraction to remove organic contaminants from wastewater can be realized by considering a typical extraction process (Figure 1). Contaminated water is extracted by a SCF in a countercurrent flow extraction column. The extract stream goes through a pressure reduction to separate the contaminant from the SCF which is recompressed and recirculated. The raffinate (decontaminated water) is also expanded to recover dissolved SCF solvent. The process would typically be at ambient temperature and pressures above 8 MPa. If a liquid solvent was used for the process two additional separation processes would have been necessary, separation of the solute from the solvent for solvent recovery and separation of residual solvent from the water stream.

Energy shortages of 1970s fueled studies on separation of alcohols from water. Therefore earlier work on SCF extraction of an aqueous medium concentrated on alcohol/water separations. Alcohols of interest were ethanol, n-propanol, i-propyl alcohol, and n-butanol and the thermodynamics of the ternary systems (alcohol/water/SCF) are well studied (*10*). SCF extraction of various other organic compounds are also reported such as acetone from water (*11*), aroma constituents of fruit and other foods (*7*), 23 organic compounds typically found in water for analytical purposes (*12*).

Stringent environmental regulations demand removal of trace organics from water making SCF extraction a viable alternative to other remediation technologies such as bioremediation, adsorption, liquid extraction, air-stripping, distillation, incineration, etc. all of which have their advantages and disadvantages. Supercritical carbon dioxide (SCCO$_2$) is the preferred SCF solvent for environmental applications since it is non-toxic, inexpensive, and readily available. In addition, it has a conveniently low critical temperature (304 K) and a moderate critical pressure (7.39 MPa).

Economic evaluation of a separation process necessitates the distribution coefficient (also called partition coefficient, equilibrium constant, and K value) of the extracted component between the phases, i.e. for SCF extraction of a solute from water the phases would be the aqueous phase and the SCF phase. Thus, studies to date concentrated on the thermodynamics of the SCF/water equilibrium. We have reported on distribution coefficients of single components such as phenol, benzene, toluene, naphthalene, and parathion (*13-15*) as well as mixtures such as a solution containing benzene, toluene, naphthalene and parathion (*15*), a phenolic mixture (*16*), and petroleum creosote (*17*). Similarly Ghonasgi et al. (*18*) and Knopf (*19*) reported on extraction of phenol, m-cresol, p-chlorophenol, and benzene both individually and as a mixture.

The thermodynamic modeling of the equilibrium starts with equating fugacities of the component distributed between the two phases

$$\hat{f}_i^{WP} = \hat{f}_i^{SCFP}$$

where the superscripts WP and SCFP refer to the water phase and the supercritical fluid phase respectively. Expressing the fugacities in terms of the mole fractions and fugacity coefficients results in the distribution coefficient

$$K_i = \frac{x_i^{SCFP}}{x_i^{WP}} = \frac{\hat{\phi}_i^{WP}}{\hat{\phi}_i^{SCFP}}$$

The fugacity coefficients $\hat{\phi}_i^j$ can be calculated from an equation of state using standard rigorous thermodynamic relationships. In our studies we have used the Peng-Robinson equation of state (*13,15*) whereas Ghonasgi et al. (*18*) used the Carnahan-Starling-DeSantis-Redlich-Kwong equation of state. It is interesting to note that hydrophilic compounds, such as phenol, have distribution coefficients in the range 0.3-2 whereas the distribution coefficients of hydrophobic compounds were two to four orders of magnitude higher, in the range 100's to 10,000's. Table I summarizes the distribution coefficients of single organic compounds in supercritical CO_2. SCF extraction, based on these high K_i values, would be feasible for extraction of hydrophobic species from water.

Effect of Entrainers in Aqueous System Extractions

Solvent power of SCFs can be increased significantly by addition of small amounts of co-solvents, called entrainers. Entrainers, usually, are polar compounds that are added to the SCF in small amounts (typically <5%) and they increase the solubility of organics in the SCF by orders of magnitude. Methanol and toluene are the most widely studied entrainers. In the literature, there are significant amount of studies reporting on the increase of the solubilities of solids in SCFs by entrainers. The increase in solubility is explained using solution theories. However, studies on the effect of entrainers on SCF extraction, i.e. the distribution coefficients, are more limited.

Roop and Akgerman (*13, 16*) showed that methanol has no effect on the distribution coefficients of phenol and a phenolic mixture proving that an entrainer that increases the solvent power does not necessarily affect the partitioning of a compound between phases. They have explained the entrainer effect in terms of multicomponent molecular interactions in calculation of thermodynamic equilibrium. Through a quaternary system equilibrium calculation, it was shown that methanol will not have any effect on phenol distribution coefficients and that benzene would increase the distribution coefficient of phenol by 50-70% and the prediction was verified experimentally. Similarly Yeo and Akgerman (*15*) determined the effect of benzene, toluene, naphthalene and parathion on each other's distribution coefficient and showed that the distribution coefficient of each component in the mixture is increased due to the presence of others as compared to single component extractions. A significant conclusion is that the extraction of all the components are enhanced in mixtures, i.e. extraction thermodynamics of mixtures is more favorable. Tables II and III summarize the enhancement of individual K_i values in a mixture for two different systems (*15, 19*).

Supercritical Extraction of Solid Environmental Media

Supercritical extraction has been demonstrated in the literature at the bench scale for extraction of organic contaminants from a variety of solid matrices. Capriel et al. (*20*)

Table I: Distribution coefficients of single organic compounds for
partitioning between water and supercritical carbon dioxide

Compound	Pressure bar	Temperature K	K	Reference
Phenol	69	298	0.44	Roop and Akgerman, (13)
	138	298	0.67	
	207	298	0.96	
	276	298	1.14	
	97	313	0.46	Ghonasgi et al., (18)
	124	313	1.27	
	172	313	1.73	
	97	323	0.34	
	110	323	0.45	Roop and Akgerman, (13)
	110	323	0.36	Ghonasgi et al., (18)
	172	323	1.66	
	207	323	0.80	Roop and Akgerman, (13)
	310	323	0.99	
m-Cresol	97	313	0.60	Ghonasgi et al., (18)
	124	313	2.20	
	165	313	3.90	
	97	323	1.00	
	124	323	1.77	
	165	323	6.02	
p-Chlorophenol	110	313	3.00	Ghonasgi et al., (18)
	138	313	4.96	
	165	313	5.38	
	97	323	0.44	
	124	323	1.32	
	152	323	2.40	
Benzene	97	313	1958	Ghonasgi et al., (18)
	124	313	3726	
	152	313	4102	
	80	318	857	Yeo and Akgerman, (15)
	90	318	1626	
	101	318	2756	
	110	323	1628	Ghonasgi et al., (18)
	138	323	2324	
	165	323	2553	
Toluene	79	318	1096	Yeo and Akgerman, (15)
	94	318	2144	
	108	318	5083	
Naphthalene	81	318	125	Yeo and Akgerman, (15)
	90	318	211	
	100	318	347	
Parathion	80	318	13.0	Yeo and Akgerman, (15)
	90	318	14.4	
	108	318	18.3	

Table II: Distribution coefficients of individual species in a mixture of phenol, m-cresol, p-chlorophenol, and benzene partitioning between water and supercritical carbon dioxide

Compound	Pressure bar	Temperature K	K	Reference
Phenol	97	313	0.51	Knopf et al., (19)
	124	313	1.15	
	152	313	1.51	
	97	323	0.32	
	124	323	0.81	
	152	323	1.27	
m-Cresol	97	313	0.63	Knopf et al., (19)
	124	313	2.66	
	152	313	3.11	
	97	323	0.98	
	124	323	1.82	
	152	323	4.90	
p-Chlorophenol	97	313	2.10	Knopf et al., (19)
	124	313	3.43	
	152	313	5.16	
	97	323	0.51	
	124	323	1.38	
	152	323	2.36	
Benzene	97	313	2020	Knopf et al., (19)
	124	313	3560	
	152	313	4220	
	97	323	960	
	124	323	1300	
	152	323	2200	

Table III: Distribution coefficients of individual species in a mixture of benzene, toluene, naphthalene, and parathion partitioning between water and supercritical carbon dioxide

Compound	Pressure bar	Temperature K	K	Reference
Benzene	81	318	911	Yeo and Akgerman, (15)
	92	318	2554	
	104	318	3730	
	82	330	444	
	97	330	810	
	110	330	1005	
Toluene	81	318	808	Yeo and Akgerman, (15)
	92	318	2568	
	104	318	5236	
	82	330	647	
	97	330	1433	
	110	330	2092	
Naphthalene	80	318	205	Yeo and Akgerman, (15)
	91	318	766	
	103	318	1088	
	81	330	65	
	97	330	98	
	110	330	291	
Parathion	81	318	22.2	Yeo and Akgerman, (15)
	92	318	29.1	
	104	318	46.3	
	82	330	6.7	
	97	330	11.8	
	110	330	15.4	

used supercritical methanol to extract bound pesticide residues from soil and plant residues. Hawthrone and Miller (*21*) extracted polycyclic aromatic hydrocarbons (PAH) from diesel soot and Tenax packing for gas chromatographic columns by supercritical carbon dioxide. Schantz and Chesler (*22*), similarly, used supercritical CO_2 to extract polychlorinated biphenyls (PCB) from sediment and PAHs from urban particulate matter. Methanol/N_2O mixtures are also used to extract PAHs from river sediments and urban particulate matter (*23*). Supercritical CO_2 is used to remove hexachlorocyclohexane, parathion, PCBs, and PAHs from Tenax packing (*24*) and polyimide based adsorbents (*25*). Supercritical CO_2 is also the solvent of choice for extraction of PAHs from various adsorbents selectively by varying the operating conditions (*26, 27*). A significant amount of work concentrated on activated carbon regeneration, such as desorption of phenol (*28*), pesticides (*29*), acetic acid and alachlor (*30*), and ethyl acetate (*31*).

The results of these studies have demonstrated at the bench scale that it is possible to extract compounds with molecular weight as high as 400 at mild conditions and selectively if desired. This conclusion constitutes the basis of supercritical chromatography. Recently we have reported on the application of chromatography theory to supercritical extraction from solid matrices (*32*).

Concerning environmental applications, Kingsley (*33*) applied subcritical and supercritical CO_2 for extraction of oil from metal fines (mill scale) and bleaching clay in the pilot scale. The process operated on a semi- batch mode and the results indicated that the recovery of extractable material depended on the solvent flow rate. It was also observed that, for the bleaching clay, the recovery was improved by a static soaking period before extraction. Brady et al. (*34*) demonstrated the ability of SCFs to extract PCBs and DDT from contaminated topsoil and subsoils using SCCO2. The effect of entrainers on the solvent power was also determined (*35*). It is demonstrated that 100% DDT extraction was possible when 5% by weight methanol was used as entrainer whereas when toluene was used as the entrainer (5% by weight) only 75 % of the DDT was recovered. Eckert et al. (*36*) studied the removal of chlorinated aromatics such as trichlorophenol (as a model compound for PCBs and dioxins) from soil. Using supercritical ethylene virtually all trichlorophenol was removed from soil.

Supercritical extraction of organic contaminants from dry solids, such as soil, sediments, adsorbents, etc., involves two different phenomena simultaneously. The organic contaminant on the solid surface can exists in two states, adsorbed state and the deposited state. The portion of the organic that is deposited as a separate phase on the solid surface is extracted by simple dissolution in the supercritical phase. On the other hand, extraction of the organic that is adsorbed on the solid phase in presence of the supercritical fluid is controlled by the adsorption/desorption equilibrium. This means that even if the organic species were present as a separate phase in mixture with the solid matrix, as soon as the supercritical fluid is introduced, a new equilibrium is established and the organic present will partition between the two phases (solid and SCF phases) as well as being present as a separate organic phase. The rule of thumb is that the amount of organic present which is in excess of the amount that can be adsorbed in presence of the supercritical fluid will be extracted by dissolution. Entrainers play a significant role in extraction of the deposited separate organic phase since the solubility of the organic in the SCF phase is significantly increased due to the entrainer, however, their effect on desorption is not that well established. In addition, SCFs have an advantage compared to conventional solvents since they penetrate the pore structure more efficiently due to their negligible surface tension and remove the organic condensed/deposited in the pores of the matrix.

Extraction of organics from a solid matrix that is also wet (contains water such as soil moisture) is more complicated since the organic now can exist in five different states, adsorbed on dry soil, dissolved in SCF, dissolved in soil moisture, adsorbed on soil which is covered by a water layer (partitioning between soil and water), and as a separate organic phase. Again, the extraction of the separate organic phase is a dissolution phenomenon. However, the extraction of the other forms involves different types of equilibrium (soil/water/SCF partitioning) combined with extraction of multiple species (water and the organic contaminant[s]).

Most important of these extraction processes is the extraction of the adsorbed species which involve solid/organic/SCF binary and ternary interactions. Thermodynamics of solids extraction by supercritical fluids is formulated in a similar manner to any adsorption process. The distribution constant of a species between the solid phase and the SCF phase is more often referred to as the adsorption equilibrium constant or the partition coefficient. The partitioning is explained in terms of an adsorption isotherm. The adsorption equilibrium constant, which is the slope of the adsorption isotherm at any concentration level, is defined as K_i^{ads} = mass adsorbed on the solid phase in (mass organic/mass solid)/concentration in the SCF phase in (moles organic/volume SCF). The concentration in the supercritical phase can also be expressed as mass concentration (mass organic/mass SCF). Hence caution should be used in stating K_i^{ads} values and units must be specified.

We have reported on the adsorption isotherms of naphthalene, phenanthrene, hexachlorobenzene, and pentachlorophenol on soil and on activated carbon in presence of supercritical CO_2. The data showed that except for pentachlorophenol, all isotherms exhibit a linear behavior on soil and the amount adsorbed increases with temperature, whereas for adsorption on carbon all isotherms were non-linear. The non-linear isotherms were fitted with the Freundlich isotherm. The amount adsorbed, however, decreases with pressure at constant temperature (*37-39*). The adsorption equilibrium constant has important thermodynamic properties. Through the van't Hoff equation, isobaric temperature dependency of adsorption equilibrium constant yields the heat of adsorption and the isothermal density dependence of the adsorption constant yields the partial molar volume of the solute in the supercritical phase.

In presence of $SCCO_2$ the amount of organic adsorbed, on soil and carbon, can be as high as 5% and 35% by weight, respectively. Hence the majority of solids extraction processes are desorption phenomena in presence of supercritical fluids and the extent of extraction (equilibrium) is governed by the thermodynamics of adsorption rather than the solubility. This fact is often overlooked in the literature. Modeling of a solids extractor involves hydrodynamic and mass transfer parameters in addition to the adsorption equilibrium and kinetics. The model used is normally a heterogeneous model that employs a dispersed flow equation in the extractor bed combined with an equation for diffusion into solid particles which includes an accumulation term for adsorption/desorption. The two equations are coupled through the external film mass transfer boundary condition and the surface accumulation term is related to the adsorption isotherm. Solution of the set of equations yield the adsorption or desorption breakthrough profiles (*40*). The parameters involved are the axial dispersion coefficient, the film mass transfer coefficient, the effective diffusivity, and the adsorption isotherm parameters. We have shown that the effective diffusivity and the adsorption parameters are the controlling parameters at normal operating conditions (*40*).

The CF Systems Corporation developed a SCF extraction process to separate and recover oils from refinery sludges and to extract hazardous organic compounds from wastewater, sludge, sediment and soil (*41*). The process uses supercritical propane on contaminated solids and $SCCO_2$ to treat wastewater. The solid feed materials are reduced in size and are slurried so that they can be pumped to the extractor. Wastewater is used directly. The process closely resembles the flow chart presented in Figure 1. Reportedly, up to 90% of the solvent is recycled in the system; the remaining 10% retains the extracted contaminants. The CF Systems process was demonstrated at pilot scale for the U.S. EPA's SITE program and shown to be capable of removing PCBs from sediments. An alternate scheme for extraction of solids as shown in Figure 2 is also proposed (*42*). This scheme eliminates the expansion of the SCF for separation of the extracted contaminants that necessitates a costly re-compression stage. Instead, the extracted contaminants are deposited on a suitable adsorbent for destruction by incineration. Costs compare favorably with other treatment techniques.

Reactions in Supercritical Fluid Solvents

Research on the use of supercritical fluids as the reaction medium in place of more conventional solvents has been receiving increasing attention. The main reactions where $SCCO_2$ are used are Diels-Alder reactions, enzymatic catalysis, polymerization, hydroformylation, hydrogenation and co-oligomerization reactions. Kaupp (*43*) reviewed the published studies on reactions in supercritical carbon dioxide. He noted that radical brominations and polymerizations, hydroformylations, CO_2 hydrogenations, catalytic additions/cycloadditions on CO_2, and enzymatic reactions have been shown to proceed as good or even better in $SCCO_2$ than in conventional solvents. Recently, Savage et al. (*44*) published an excellent review on reactions in supercritical fluids with an extensive section on the fundamentals and applications of supercritical water oxidation. Only some of the major work related to pollution prevention by having the reaction in a supercritical fluid solvent rather than an organic solvent are cited below and the work on supercritical water oxidation is not included since extensive reviews are available (*44*).

The Diels-Alder reaction is a non-catalytic reaction which involves the addition of an alkene or alkyne to a diene. Paulaitis and Alexander (*45*) studied the reaction of maleic anhydride with isoprene in $SCCO_2$ at conditions near the critical point of CO_2. They showed that the reaction rate in $SCCO_2$ is similar to the rates in organic solvents and that the reaction rate increases with pressure near the critical point due to large negative partial molar volumes. The Diels-Alder reaction is also studied in $SCCO_2$ by Isaacs and Keating (*46*) who investigated the reaction between p-benzoquinone and cyclopenta-1,3-diene in $SCCO_2$ and Ikushima et al. (*47*) who studied reaction of isoprene and methyl acrylate. The purpose of these studies was to investigate the effect of the presence of a supercritical phase on the reaction chemistry. Specifically, the interest focused on the rate increase in supercritical fluids due to a clustering effect, i.e. large

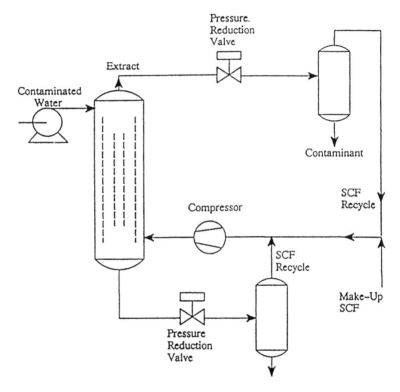

Figure 1. Schematic representation of a typical supercritical extraction process
(Reproduced with permission from reference 17. Copyright 1991 CRC.)

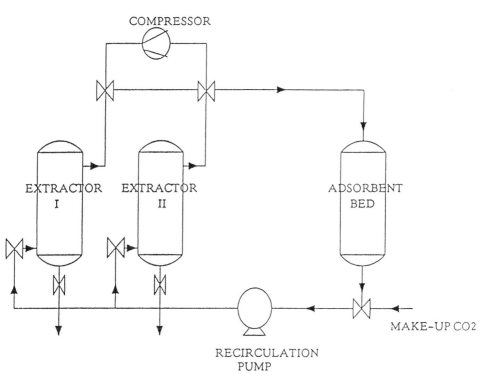

Figure 2. Schematic representation of a supercritical extraction process with subsequent adsorption of the extracted contaminants for soil remediation. (Reproduced with permission from reference 422. Copyright 1994 American Institute of Chemical Engineers.)

negative partial molar volumes. Recently Randolph et al. (*48*) showed that solute-solute correlations contribute to reaction rates at low density but not at high density where the reaction is in the transport limited regime.

Tanko and Blackert (*49*) studied the bromination of toluene and ethylbenzene in supercritical carbon dioxide with photochemical initiation. They reported high yields and rates similar to that observed in conventional organic solvents. The side reactions were, however, minimized, indicating the advantage of using a supercritical fluid solvent. The products were predominantly benzyl bromide and 1-bromo-1-phenylethane forming from the reactivity of the benzylic secondary C-H bond.

$SCCO_2$ also has been proven to be a feasible reaction medium for enzymatic reactions under some reaction conditions. Indeed it was shown that supercritical media, compared with aqueous media, achieve good enzyme stability, increased solubility of hydrophobic compounds, and easy enzyme recovery as they are usually insoluble in in these media (*50-52*). Hammond et al (*53*) investigated the catalytic activity of polyphenol oxidase in $SCCO_2$ at 309 K and 34.45 MPa, p-cresol and p-chlorophenol were readily oxidized to their corresponding o-benzoquinones. Conversions of 70-80% were achieved for oxidation of p-cresol whereas the conversion for p-chlorophenol was 27%. Randolph et al. (*54*) investigated the reaction of cholestrol to form cholest-4-en-one in $SCCO_2$ with cholestrol oxidase. A limitation of this work was the low solubility of the reactants and products in $SCCO_2$. They also performed a more detailed investigation of the same reaction in a flow reactor with a 9:1 CO_2/O_2 mixture (*55*). Most studies of enzymatic reactions in supercritical fluids were of short duration and in batch reactors, but van Eijs et al. (*56*) have performed a continuous transesterification reaction in $SCCO_2$ in a packed bed of enzymes. No decrease in productivity was observed during a 12 day period of operation. These studies, and many others in the literature, indicate that enzymatic reactions (esterifications, transesterifications, interesterifications, and oxidations) in $SCCO_2$ are possible, however, no clear advantage over using other organic solvents is indicated as far as reaction rates and selectivities are concerned (*57*). On the other hand, in addition to being environmentally benign, supercritical fluids have the noted advantage in separation of the products from the reaction media.

A patent by Kramer and Leder (*58*) describes a C_4-C_{12} normal or isoparaffinic hydrocarbon isomerization process employing $SCCO_2$. At the temperatures and pressures used (up to 80°C and up to 170 bar) the preferred Friedel Crafts catalyst, aluminum bromide, was soluble and used homogeneously in the concentration range 0.1-0.5 M. Addition of soluble gaseous hydrogen to the medium aided selectivity to isomerization over cracking. Conversions were similar to heterogeneous liquid-solid reactions, but with greater selectivity for the isomerization products. The advantage of the process was removal of reaction products and/or spent catalyst by a temperature and/or pressure swing. Poliakoff et al. (*59*) investigated the formation of metal complexes of gases (photolysis of metal carbonyls in the presence of hydrogen or nitrogen) dissolved in a supercritical medium and have shown that the process is superior compared to the conventional technique of synthesis in liquified noble gases. Baumgartner (*60*) achieved *tert*-butyl hydroperoxide synthesis in supercritical isobutane containing very small amounts of oxygen.

DeSimone and colleagues showed that carbon dioxide is an excellent medium to perform free-radical reactions such as polymerization (*61-64*). They studied the synthesis of fluoropolymers and found that the rate of disappearance of the initiator (azobisisobutyronitrile) was 2.5 times higher in benzene than in CO_2. Pernecker and Kennedy (*65-67*) studied the carbocationic polymerization of isobutylene in $SCCO_2$ by the use of Friedel-Crafts acid-based initiating systems and noted that the molecular weights of the obtained polyisobutylenes (PIB) were an order of magnitude higher than PIBs obtained using conventional solvents under identical experimental conditions. The most significant aspect of their study was the ability to achieve significant extents of polymerization at 32.5°C (M_n~2000, M_w/M_n~2). Normally these polymerization reactions are conducted at low temperatures in conventional solvents necessitating cooling of the reactor system and at these temperatures only very low molecular weight oligomers (~tetramers) are obtained. Adamsky and Beckman (*68*) reported high yields for the inverse emulsion polymerization of acrylamide in $SCCO_2$ and obtained polymers of high molecular weight. A patent (*69*) describes a process for preparing polymers having a molecular weight below 5,000 by polymerization in supercritical carbon dioxide. The proposed conditions are, however very extreme with temperatures of at least 200°C and pressure above 3,500 psi. Earlier, Fukui et al. (*70*) proposed a novel method for polymerizing a vinyl compound in the presence of $SCCO_2$.

In some instances, carbon dioxide can be used simultaneously as a medium and as a reactant. Jessop et al. (*71, 72*) studied the hydrogenation of $SCCO_2$ to formic acid and dimethylformamide using ruthenium-based catalysts and measured a very high catalytic activity and high rates of formation of the products much higher in $SCCO_2$ than in water or in other organic solvents.

Supercritical fluids are also used as the reaction media in heterogeneous catalysis. Recently studies on catalytic partial oxidation of toluene in $SCCO_2$ over a supported CoO catalyst (*73*) and complete oxidation of toluene and an aromatic-aliphatic hydrocarbon mixture in $SCCO_2$ over a Pt/Al_2O_3 catalyst (*74*) were reported. Pang et al. (*74*) observed complete combustion at reaction temperatures over 523 K, but at lower temperatures partial oxidation products were detected. On CoO catalysts calcined and/or reduced at different conditions, Dooley and Knopf (*73*) report apparent activation energies in the range of 5.1-38 kcal/mol. We studied oxidation of toluene, tetralin, decane, ethanol, acetaldehyde, and butanone in $SCCO_2$ over a Pt/alumina or Pt/titania catalyst in relation to an extraction/destruction process for soil remediation and have shown that the oxidation products consist of CO_2 and H_2O (*75, 76*). Fischer-Tropsch synthesis is also studied in supercritical propane and hexane (*77-80*). Although the Schultz-Flory distribution was very similar to the gas phase Fischer-Tropsch reaction, a significant increase in α-olefins is observed which is explained in terms of the pore cleaning characteristics of supercritical fluids. Similar pore cleaning effect of supercritical reaction media is also observed for the hexene isomerization reaction (*81,82*).

Recently we studied the homogeneously catalyzed propylene hydroformylation reaction in supercritical carbon dioxide at different temperatures and pressures using $Co_2(CO)_8$ catalyst. Figure 3 shows propylene conversion as a function of time at different pressures at 88°C temperature. As can be seen from the figure, at a given reaction time conversion increases with increasing pressure. In this study we used the same amount of reactants (propylene, hydrogen, carbon monoxide) and the same amount of catalyst in a constant volume reactor. Hence the initial concentrations are the same in each run.

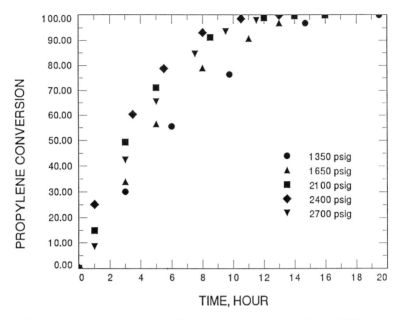

Figure 3. Propylene conversion in hydroformylation reaction at 88°C and different total pressures.

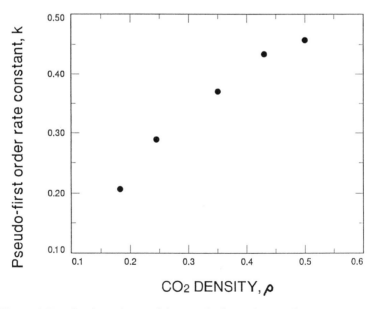

Figure 4. Density dependence of the pseudo first order reaction rate constant at 88°C.

Therefore, we believe that the observed effect is due to the pressure (the amount of carbon dioxide) in the reactor volume. This effect cannot be due to carbon dioxide playing a role in the reaction since the net effect of that would be a reduction in rate with pressure, in contradiction to the observed effect. In addition, the amount of catalyst used, in all cases, is less than the amount corresponding to the solubility in the supercritical mixture, hence the effect cannot be due to the catalyst concentration as well. Figure 4 shows how the pseudo first order rate constant k varies with density. The Arrhenius dependency of the rate constant measured at 2400 psia is presented in Figure 5, showing an activation energy of 30 kcal/mol, which is comparable to the 27-35 kcal/mol measured in organic solvents (*83*). The reaction selectivities measured, at the termination of the reaction, is about 80% for the linear aldehyde vs the branched aldehyde and seems to be independent of the pressure and the temperature. This is one of the very first detailed studies on a homogeneously catalyzed reaction in a supercritical fluid at conversions up to 100%. A different, and potentially beneficial phenomena is taking place. We believe that homogeneously catalyzed reactions in supercritical fluids would enable better reaction rate and selectivity control by pressure.

Adsorptive Separations by Frontal Analysis Supercritical Chromatography

Separation of thermally labile and structurally very similar compounds usually employ organic solvents that are coming under close scrutiny. Of special interest is separation of high value, low vapor pressure isomers and, especially, optical isomers (enantiomers) in fine chemicals and pharmaceutical industries where toxic organic liquid solvents are used as the extraction/separation media. Supercritical fluids offer a novel separation technique. The mixture to be separated, such as a mixture of enantiomers, would be dissolved in a supercritical solvent (such as supercritical carbon dioxide) and passed over an adsorption column. The desired compound is either concentrated in the column by adsorbing on the adsorbate or concentrated in the effluent. The column is then regenerated by pure solvent. The technique, called frontal analysis supercritical fluid chromatography, is a separation technique based on different adsorptive interactions of the species involved with the adsorbent combined with adsorption column hydrodynamic and mass transfer characteristics. Thus the system consists of an adsorbent material (solid phase), the supercritical fluid (mobile phase), and components A and B (solutes) which are partitioning between the two phases. Use of supercritical fluids not only eliminates consumption of toxic organic solvents, but results in pure species without any solvent residue. This adsorptive separation process involves multicomponent adsorption-desorption phenomena using the supercritical fluid as the mobile phase. The thermodynamics of such multicomponent adsorption processes are governed by multicomponent solubilities in supercritical fluids and multicomponent adsorption isotherms on solid matrices from supercritical fluids. The kinetics, on the other hand, are governed by the adsorbent particle size, column hydrodynamics, multicomponent intraparticle diffusion and interparticle mass transfer.

Separation by adsorption can be performed in two ways, by size exclusion due to the molecular size difference of the components (size exclusion chromatography), and by adsorptive interactions due to different adsorptive strengths of the components, the best example being chromatographic separations. Sakanishi et al., (*84*), reported on selective separation of indole from methylnaphthalene oil. The oil (containing ~3% indole) was dissolved in $SCCO_2$ and Amberlite IRA-904 resin was used as the adsorbent. The indole breakthrough from the column was measured and the resin regenerated by methanol. The methanol extract of the resin was 98+% indole. The separation was due to selective adsorption of indole on the anion exchange resin. Indole has a slightly acidic N-H group which can be exchanged by its anion form with the corresponding anion of the resin.

Work on enantiomer separations focus on analytical aspects of supercritical chromatography. The low temperatures (leading to a more ordered stationary phase) and high diffusion rates associated with supercritical fluids provide special benefits for enantiomeric separations by supercritical fluid chromatography. Chromatographic chiral separations, particularly HPLC, has been rapidly advanced in the last 15 years and many chiral stationary phases have been developed (85). Chiral separations by supercritical chromatography employs mainly these chiral stationary phases, which have better selectivity for adsorption of one of the enantiomers. There is a significant amount of work in the literature on chiral separations of enantiomers (86, 87). Fuchs et al. (88) describe a pilot plant scale separation of phosphine oxide enantiomers. Optical purity obtained in the separation was better than 95%. Siret et al. (89) separated enantiometric β-blockers by packed column supercritical fluid chromatography. They postulated that CO_2, is not just a passive solvent, but plays an active role in the separation by forming a complex with the 1,2-amino alcohol moiety of the drug which enhances enantioselectivity of the separation. Bargmann-Leyder et al. (90) used both normal liquid phase chromatography and supercritical fluid chromatography for separation of β-blockers on ChyRoSine-A and showed that resolution is much better using supercritical fluid chromatography. They suggest that the solvating effect of carbon dioxide induces a change of the propranolol conformations that is geometrically favorable to the chiral discrimination. Sakaki and Hirata (91) separated secondary alcohol enantiomers by derivatizing the alcohols to form diasteriomers. Hara et al. (92) achieved optical resolution of α-amino acid derivatives using carbon dioxide supercritical chromatography on n-formyl-valine bonded silica gel chiral stationary phase. Similarly, Mourier et al. (93) separated phosphine oxide enantiomers and Steuer et al. (94) separated 1,2-aminoalcohol enantiomers on chiral stationary phases. These studies prove that enantiomers can be separated by using a proper adsorbent. However, the focus had been on separation for identification purposes rather than separation for purification and recovery purposes.

Frontal analysis chromatography has been developed as an analysis technique and is widely applied for preparative separations. However, extension of the frontal analysis chromatography principles to large scale adsorptive separations is just emerging as a viable option for separation of structurally very similar compounds or enantiomers.

Selective Extractions by Supercritical Fluids

Supercritical carbon dioxide extraction of natural products has two major advantages over traditional steam distillation and solvent extraction processes. First, it can be operated at lower temperatures to avoid thermal degradation of thermally labile precursors, flavors or fragrances. Second, no solvent residue will remain after separation by pressure reduction. Extensive studies have been made on $SCCO_2$ extraction of essential oils. Calame and Steiner (*95*) studied the extraction of essential oil from lilac, lemon peel, black pepper, and almonds. Schultz and Randall (*7*) described the selective aroma extraction from apple, pear, orange juices, orange pieces, and roasted-ground coffee using liquid carbon dioxide. Sankar (*96*) reported extraction of volatile oils from pepper (Piper nigrum). Extraction of flavors from juniper berries using $SCCO_2$ was reported by Nykanen et al. (*97*). $SCCO_2$ is used for the extraction of various other plants to obtain aroma concentrates and essential oils (*98-101*). Temelli et al. (*102, 103*) discussed the application of $SCCO_2$ extraction to remove terpenese from cold-pressed Valencia orange oil. All these studies concentrated on total oil extraction rather than the selectivity of the extraction process for the desired components of the essential oils.

As in all separation processes, phase equilibrium data (i.e. solubilities of the essential oil components in the supercritical solvent) are needed for supercritical extraction. Francis (*104*) gave the solubility data for 261 chemicals, including some components of essential oils, in liquid CO_2. Stahl and Gerard (*105*) studied and reported on the solubilities of pure essential oil components in $SCCO_2$. Coppella and Barton (*99*) investigated vapor-liquid equilibrium at 30-40°C and 4-9 MPa for CO_2-limonene system. Temelli et al. (*102, 103, 106*) studied the phase equilibrium for $SCCO_2$ extraction of terpenes from cold-pressed orange oil. However, thermodynamic studies on $SCCO_2$ extraction of essential oils are not extensively reported. In addition, the chemical equilibrium data for extraction of essential oils from solid phase species, such as apple powder, lemon peel, etc., by SCCO2 are not available.

Although supercritical extraction is used, at least at laboratory scale, for extraction of natural substances, there are no studies on extraction of certain compounds selectively. We have recently studied supercritical carbon dioxide extraction of lavender flowers (*Lavandula Stoechas* subsp. *Cariensis Boiss)* both in batch equilibrium cells and in flow system. Figure 6 is a plot of total oil yield as a function of temperature and pressure. More interesting, however, is the selective extractability of four components pinene, fenchore, camphor and t-muurolol shown in Figure 7. Using these batch equilibrium studies we performed a flow experiment at the optimum extraction conditions of 35°C and 80 bar for the extraction of these components. Figure 8 shows that at the onset of the breakthrough fenchone and camphor constitute almost 85% of the effluent, the remainder being 14 % in only three components and a trace amount of oils. This figure demonstrates that selective extraction of the desired components is possible by properly setting the extraction conditions which can be determined from batch studies on thermodynamics (*107, 108*). The components can be further separated using the adsorptive separation process explained above.

Conclusions

In summary, supercritical fluids are finding extensive application in environmental remediation and restoration as well as in emerging applications of pollution prevention. Technologies have been developed and commercialized for extraction of contaminants from aqueous and solid environmental matrices. Use of supercritical fluids as the reaction and separations media to replace organic solvents is rapidly emerging as a viable alternative. In addition to solvent replacement, and more important, supercritical fluids

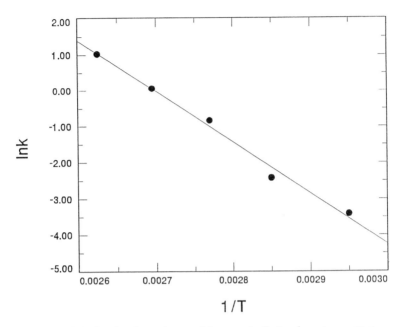

Figure 5. Arrhenius dependency of the pseudo first order rate constant.

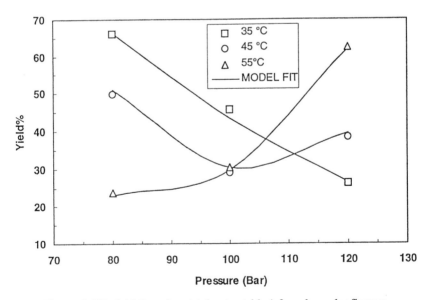

Figure 6. Oil yield (based on total extractables) from lavender flowers.

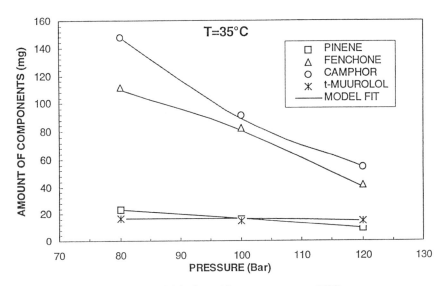

Figure 7. Yield of specific components at 35°C.

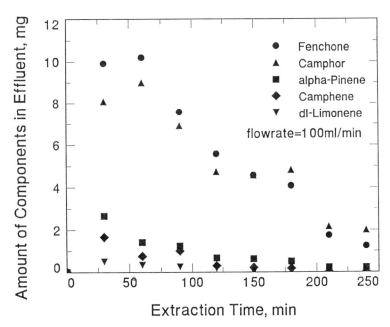

Figure 8. Breakthrough profiles for specific components of lavender oil at 35°C and 80 bar.

have advantageous properties which make their use desirable as reaction and separation media. The most important property is the control of solvency power with density resulting in selective separations and control of reaction extent and selectivity.

Literature Cited

1. Wilson, R. E.; Keith, P. C.; Haylett, R. E. *Ind. Eng. Chem.* **1936**, *28*, 1065.
2. Elgin, J. C.; Weinstock, J. J. *J. Chem. Eng. Data*, **1959**, *4*, 3.
3. Wilhelm, A.; Hedden, K. In: *Proc. Int. Conf. Coal Sci.*, Pittsburgh, 1983, p. 6.
4. Zosel, K. *Angew. Chem. Int. Ed., Engl.*, **1978**, *17*, 702.
5. Vitzthum, O. G.; Hubert, P. *U.S. Patent 3,879,569*, 1975.
6. Hubert, P.; Vitzthum, *Angew. Chem. Int. Ed., Engl.*, **1978**, *17*, 710.
7. Schultz, W. G.; Randall, J. M. *Food Technol.* **1970**, *24*, 94.
8. McHugh, M. A.; Krukonis, V. J. *Supercritical Fluid Processing: Principles and Practice;* Butterworth, Stoneham, MA, 1986.
9. Kuk, M. S.; Montagna, J. In: *Chemical Engineering at Supercritical Conditions*, Paulaitis, M. E.; Penninger, J. M. L.; Gray, R. D., Jr.; Davidson, P. (Eds.). Ann Arbor Science, Ann Arbor, MI, 1983.
10. Paulaitis, M. E.; Kander, R. G.; DiAndreth, J. R. *Ber. Bunsenges. Phys. Chem.*, **1984**, *88*, 869.
11. Panagiotopoulos, A. Z.; Reid, R. C. *ACS Prep. Div. Fuel Chem.*, **1985**, *30*, 46.
12. Ehntholt, D. J.; Thrun, K.; Eppig, C. *Int. J. Environ. Anal. Chem.*, **1983**, *13*, 219.
13. Roop, R. K.; Akgerman, A. *Ind. Eng. Chem. Res.*, **1989**, *28*, 1542.
14. Roop, R. K.; Akgerman, A.; Dexter, B. J.; Irvin, T. R. *J. Supercrit. Fluids*, **1989**, *2*, 51.
15. Yeo, S.-D.; Akgerman, A. *AIChE J.*, **1990**, *36*, 1743.
16. Roop, R. K.; Akgerman, A. *J. Chem. Eng. Data*, **1990**, *35*, 257.
17. Akgerman, A.; Roop, R. K.; Hess, R. K.; Yeo, S.-D. In: *Supercritical Fluid Technology: Reviews in Modern Theory and Application*, Bruno, T. J.; Ely, J. F. (Eds.)., CRC Press, Boca Raton, 1991.
18. Ghonasgi, D.; Gupta, S.; Dooley, K. M.; Knopf, F. C. *AIChE J.*, **1991**, *37*, 944.
19. Knopf, F. C., Louisiana State University, personal communication, 1992.
20. Capriel, P.; Haisch, A.; Khan, S. U. *J. Agric. Food Chem.*, **1986**, *34*, 70.
21. Hawthrone, S. B.; Miller, D. J. *J. Chromatogr. Sci.*, **1986**, *24*, 258..
22. Schantz, M. M.; Chesler, S. N. *J. Chromatogr.*, **1986**, *363*, 397.
23. Hawthrone, S. B.; Miller, D. J. *Anal. Chem.*, **1987**, *59*, 1705.
24. Raymer, J. H.; Pellizzari, E. D. *Anal. Chem.*, **1987**, *59*, 1043.
25. Raymer, J. H.; Pellizzari, E. D.; Cooper, S. D. *Anal. Chem.*, **1987**, *59*, 2069.
26. Wright, B. W.; Wright, C. W.; Gale, R. W.; Smith, R. D. *Anal. Chem.*, **1987**, *59*, 38.
27. Wright, B. W.; Frye, S. R.; McMinn, D. G.; Smith, R. D. *Anal. Chem.*, **1987**, *59*, 640.
28. Kander, R. G.; Paulaitis, M. E. In: *Chemical Engineering at Supercritical Conditions.* Paulaitis, M. E.; Penninger, J. M. L.; Gray, R. D., Jr.; Davidson, P. (Eds.), Ann Arbor Science, Ann Arbor, MI, 1983.
29. DeFilippi, R. P.; Krukonis, V. J.; Robey, R. J.; Modell, M. *EPA-600/2-80-054*, U.S. Environmental Protection Agency, Washington, D. C., 1980.

30. Picht, R. D.; Dillman, T. R.; Burke, D. J. *AIChE Symp. Ser.*, **1982**, *78*, 136.
31. Tan, C.; Liou, D. *Ind. Eng. Chem. Res.*, **1988**, *27*, 988.
32. Erkey, C.; Akgerman, A. *AIChE J.*, **1990**, *36*, 1715.
33. Kingsley, G. S. *EPA/600/2-85/081*, U. S. Environmental Protection Agency, Washington, D.C., 1985
34. Brady, B. O.; Kao, C. C.; Dooley, K. M.; Knopf, F. C.; Gambrell, R. P. *Ind. Eng. Chem. Res.*, **1987**, *26*, 261.
35. Dooley, K. M.; Kao, C.; Gambrell, R. P.; Knopf, F. C. *Ind. Eng. Chem. Res.*, **1987**, *26*, 2058.
36. Eckert, C. A.; Van Alsten, J. G.; Stoicos, T. *Environ. Sci. Technol.*, **1986**, *20*, 319.
37. Hess, R. K.; Erkey, C.; Akgerman, A. *J. Supercrit. Fluids*, **1991**, *4*, 47.
38. Erkey, C.; Madras, G.; Orejuela, M.; Akgerman, A. *Environ. Sci. & Technol.*, **1993**, *27*, 1225.
39. Madras, G.; Erkey, C.; Akgerman, A. *Ind. Eng. Chem. Res.*, **1993**, *32*, 1163.
40. Madras, G.; Thibaud, C.; Erkey, C.; Akgerman, A. *AIChEJ*, **1994**, *40*, 777.
41. Hall, D. W.; Sandrin, J. A.; McBride, R. E. *Environ. Progr.*, **1990**, *9*, 98.
42. Madras, G.; Erkey, C.; Akgerman, A. *Environ. Progr*, **1994**, *13*, 45.
43. Kaupp, G.*Angew. Chem. Int. Ed., Engl.*, **1994**, *33*, 1452.
44. Savage, P. E.; Gopalan, S.; Mizan, T. I.; Martino, C. J.; Brock, E. E. *AIChE J.*, **1995**, *41*, 1723.
45. Paulaitis, M. E.; Alexander, G. C. *Pure & Appl. Chem.*, **1987**, *59*, 61.
46. Isaacs, N. S.; Keating, N. *J. Chem. Soc., Chem. Commun.*, **1992**, 876
47. Ikushima, Y.; Saito, N.; Arai, M. *J. Phys. Chem.*, **1992**, *96*, 2293.
48. Randolph, T. W.; O'Brien, J. A.; Ganapathy, S. *J. Phys. Chem.*, **1994**, *98*, 4173.
49. Tanko, J. M.; Blackert, J. F. *Science*, **1994**, *263*, 203.
50. Aaltonen, O.; Rantakyl, M. *Chemtech*, pp. 240-248, April 1991.
51. Nakamura, K., *Tibtech.*, **1990**, *8*, 288.
52.Randolph, T. W.; Blanch, H. W.; Clarck, D. S. In: *Biocatalysis for Industry*, Dordick, J. S., Ed., Plenum Press, New York, 1991, pp. 219-237.
53. Hammond, D.A.; Kerel, M.; Klibanov, A.M. *Appl. Biochem. Biotech.*, **1985**,*11*, 393.
54. Randolph, T. W.; Blanch, H. W.; Prausnitz, J. M.; Wilke, C. R. *Biotech. Letters,* **1985**, *7*, 325.
55. Randolph, T. W.; Clark, D. S.; Blanch, H. W.; Prausnitz, J. M. *Science*, **1988**, *238*, 61.
56. van Eijs, A. M. M.; de Jung, J. P. L.; Doddema, H. J.; Lindeboom, D. R. In: *Proc. Intl. Symp. Supercritical Fluids in Nice, France*, M. Perrut, Ed., Institut National Polytechnique de Lorraine, France 1988.
57. Marty, A.; Combes, D; Condoret, J.-S. *Biotech. & Bioeng.*, **1994**, *43*, 497.
58. Kramer, G. M.; Leder, F. *U. S. Patent 3,880,945*, **1975**.
59. Poliakoff, M.; Howdle, S. M.; Healy, M. A.; Whalley, J. M. In *Proc. Intl. Symp. SupercriticalFluids in Nice, France*, M. Perrut, Ed., Institut National Polytechnique de Lorraine, France 1988.
60. Baumgartner., H. J., *U. S. Patent 4,408,082*, **1983**.
61. DeSimone, J. M.; Guan, Z.; Elsbernd, C. S. *Science*, **1992**, *257*, 945.

62. Guan, Z.; Combes, J. R.; Menceloglu, Y. Z.; DeSimone, J. M. *Macromolecules*, **1993**, *26*, 2663.

63. Combes, J. R.; Guan, Z.; DeSimone, J. M. *Macromolecules*, **1994**, *27*, 865.

64. Romack, T. J.; Combes, J. R.; DeSimone, J. M. *Macromolecules*, **1995**, *28*, 1724.

65. Pernecker, T.; Kennedy, J. P. *Polymer Bull.*, **1994**, *32*, 537.

66. Pernecker, T.; Kennedy, J. P. *Polymer Bull.*, **1994**, *33,* 13.

67. Pernecker, T.; Kennedy, J. P. *Polymer Bull.*, **1994**, *33*, 259.

68. Adamsky, F. A.; Beckman, E. J. *Macromolecules*, **1994**, *27*, 312.

69. Dada, A. E.; Lau, W.; Merrit, R. F.; Paik, H.; Swift, G. *U. S. Patent 5,328,972*, **1994**.

70. Fukui, K.; Kagiya, T.; Yokota, H.; Toriuchi, Y.; Fujii, K. *U. S. Patent 3,522,228*, **1970**.

71. Jessop, P. G.; Hsiao, Y.; Ikariya, T.; Noyori, R. *J. Am. Chem. Soc.*, **1994**, *116*, 8851.

72. Jessop, P. G.; Ikariya, T.; Noyori, R. *Nature*, **1994**, *368*, 231.

73. Dooley, K. M.; Knopf, F. C. *Ind. Eng. Chem. Res.*, **1987**, *26*, 1910.

74. Pang, T. H.; Ye, M.; Knopf, F. C.; Dooley, K. M. *Chem. Eng. Commun.*, **1991**, *110*, 85.

75. Zhou, L.-B.; Erkey, C.; Akgerman, A. *AIChE J.*, **1995**, *41*, 2122.

76. Zhou, L.-B.; Akgerman, A. *Ind. Eng. Chem. Res.*, **1995**, *34*, 1588.

77. Lang, X.; Akgerman, A.; Bukur, D. B. *Ind. Eng. Chem. Res.*, **1995**, *34*, 72.

78. Yokota, K.; Fujimoto, K. *Fuel*, **1989**, *68*, 255.

79. Yokota, K.; Fujimoto, K. *Ind. Eng. Chem. Res.*, **1991**, *30*, 95.

80. Fan, L.; Yokota, K.; Fujimoto, K. *AIChE J.*, **1992**, *38*, 1639.

81. Tilscher, H.; Wolf, H.; Schelchshorn, J. *Ber. Bunsenges. Phys. Chem.*, **1984**, *88*, 897.

82. Saim, S.; Ginosar, D. M.; Subramaniam, B. In: *Supercritical Fluid Science and Technology*, Johnston, K. P., Ed., ACS Symp. Ser., No. 406, p. 301, **1989**.

83. Pino, P.; Piacenti, F.; Bianchi, M. In: *Organic Synthesis via Metal Carbonyls, V.2*, Wender, I.; Pino, P., Editors, John Wiley, New York, 1977.

84. Sakanishi, K.; Obata, H.; Mochida, I.; Sakaki, T. *Ind. Eng. Chem. Res.*, **1996**, *35*, 335.

85. Kaida, Y.; Okamoto, Y. In: *Fractionation by Packed Column SFC and SFE - Principles and Applications*, Saito, M.; Yamauchi, Y.; Okuyama, T. , Eds., pp. 215-229, VCH Publishers, New York, 1994.

86. Chester, T. L.; Pinkston, J. D.; Raynie, D. E. *Anal. Chem.*, **1994**, *66*, 106R.

87. Dean, J. R.; Editor, *Applications of Supercritical Fluids in Industrial Analysis*, Chapman and Hall Publishers, Glasgow, England, 1993.

88. Fuchs, G.; Doguet, L.; Barth, L.; Perrut, M. *J. Chromatogr.*, **1992**, *623*, 329.

89. Siret, L.; Bargmann, N.; Tambute, A.; Caude, M. *Chiralty*, **1992**, *4*, 252.

90. Bargmann-Leyder, N.; Sella, C.; Bauer, D.; Tambute, A.; Caude, M. *Anal. Chem.*, **1995**, *67*, 952.

91.Sakaki, K.; Hirata, H. *J. Chromatogr.*, **1991**, *585*, 117.

92. Hara, S.; Dobashi, A.; Hondo, T.; Saito, M.; Senda, M. *J. High Resolution Chromatog. & Chromatog. Commun.*, **1986**, *9*, 249.

93. Mourier, P. A.; Eliot, E.; Caude, M. H.; Rosset, R. H.; Tambute, A. G. *Anal. Chem.*, **1985**, *57,* 2819.

94. Steuer, W.; Schindler, M.; Schill, G.; Erni, F. *J. Chromatog.*, **1988**, *447*, 287.

95. Calame, J. P.; Steiner, R. *Chemical Industry*, p. 399, June 1982.
96. Sankar, K. U. *J. Sci. Food Agric.*, **1989**, *48*, 483.
97. Nykanen, I.; Nykanen, L.; M. Alkio, *Flavour Science and Technology*, **1990**, *6*, 217.
98. Moyler, D. A., *Perfumer & Flavorist*, **1984**, *9*, 109.
99. Copella, S. J.; Barton, P. *Prep. Paper Am. Chem. Soc. Div. Fuel Chem.* **1985**, *30*, 195.
100. Tateo, F.; Chizzini, F. *J. of Ess. Oil Res.*, **1989**, *1*, 165.
101. Tateo, F.; Verderio, E. *J. Ess. Oil Res.*, **1989**, *1*, 97.
102. Temelli, F.; Braddock, R. J.; Chen, C. S. In: *Supercritical Fluid Extraction and Chromatography: Techniques and Applications*, Charpentier, B.A.; Sevenants M. R., Eds., *Am. Chem. Soc. Symp. Ser. 366*, ACS, Washington, DC, pp. 110-126, 1988.
103. Temelli, F.; Chen, C. S.; Braddock, R. J. *Food Technol.*, **1988**, *42*, 145.
104. Francis, A. *Ind. Eng. Chem.*, **1955**, *47*, 230.
105. Stahl, E.; Gerard, D. *Perfum. Flavor.*, **1985**, *10, 29*.
106. Temelli, F.; O'Connell, J. P.; Chen, C. S.; Braddock, R. J. *Ind. Eng. Chem. Res.*, **1990**, *29*, 618.
107. Akgun, N.; Akgun, M.; Dincer, S.; Akgerman, A. *J. Agric. Food Chem.*, in review 1996.
108. Akgun, N.; Akgun, M.; Dincer, S.; Akgerman, A. *J. Agric. Food Chem.*, in review 1996.

Chapter 16

Mass Transfer and Chemical Reaction During Catalytic Supercritical Water Oxidation of Pyridine

Sudhir N.V.K. Aki and Martin A. Abraham[1]

Department of Chemical Engineering, University of Tulsa, Tulsa, OK 74104

Catalytic supercritical water oxidation has been recently demonstrated as an effective means of destroying aqueous organic wastes. By using a catalyst, the reaction temperature can be lowered to around 400°C without sacrificing the destruction efficiency. MnO_2/CeO_2 catalyst was used to enhance the conversion of pyridine during SCWO. The reaction resulted in higher yields of carbon dioxide and nitrous oxide than observed without catalyst addition. NOx was not produced in any experiments. Simple power law kinetics were used to describe the experimental results. Experiments were also conducted to study the external and internal mass transfer limitations.

Supercritical water oxidation (SCWO) has been developed to destroy organic compounds by oxidation in water at temperatures and pressures above its critical point, which is approximately 374°C and 22.1 MPa (218 atm). This process is generally conducted at 400 to 650°C and about 24 to 35 MPa, and is capable of achieving a high destruction efficiency. There is no formation of NO_x or SO_2 and *in situ* neutralization of the produced acid gases can be achieved (1).

Supercritical fluids assume gas like viscosity and diffusivity and liquid like densities. The dielectric constant of water at temperatures and pressures of interest in SCWO drops to values similar to non-polar organic compounds at ambient conditions (2), the ion product decreases below 10^{-20} mol/kg (3), and hydrogen bonding is notably less extensive (4). Gases and hydrocarbons are completely soluble in supercritical water (5), whereas inorganic salts that are soluble in sub-critical water become insoluble in supercritical water (6). Thus, SCWO can be achieved in a single phase without mass transfer limitations.

Previous work on SCWO of nitrogen containing compounds, and in particular pyridine, is limited. Crain et al., (7) studied the SCWO of pyridine in a temperature range

---

[1]Current address: Department of Chemical Engineering, University of Toledo, Toledo, OH 43606–3390

of 425 to 527°C. They found the conversion to increase from 0.03 at 426°C to 0.68 at 527°C at a residence time of 10s. In these experiments, the oxygen to pyridine molar ratio was varied between 1 and 2. These experiments were conducted in a flow reactor system. Wightman (8) also studied SCWO of pyridine in a flow system. He reported a conversion of around 16% at 400°C and a pressure of 6000 psia. In these experiments the residence time was fixed at 29 s and an oxygen concentration of 100% excess was used.

Experimental Procedure

All experiments were conducted in a small scale batch reactor, comprised of a 1/2 inch stainless steel Gyrolock cap and nut. This was connected to a high pressure taper seal valve (1/16 inch tubing size), HiP 15-12AF1, by means of Gyrolock reducers. The total volume of the reactor was 3 mL.

The required feed composition was calculated by means of the Peng-Robinson Equation of state. Catalyst, pyridine and water were added to the cap and their weights measured on a Sartorius Analytic balance, accurate up to 0.0001 g. The reactor was then sealed, connected to an air cylinder, and the required amount of air was added. The air pressure was measured by means of a 3-D Instruments Pressure Test Gauge.

The catalyst used for the catalytic supercritical water oxidation of pyridine was MnO_2/CeO_2. The catalyst was purchased from Carus Chemical. Catalysts were crushed and sieved to provide catalyst particles of the desired size. No other catalyst modifications were undertaken. All other chemicals were purchased from Aldrich Chemical Company and used as received.

The reaction was conducted in a Techne fluidized sand bath. The sand bath was heated to the required temperature, with continuous flow of air. The temperature was measured and controlled (±2°C) by means of a Tecam TC4D temperature controller. The reactors were immersed into the sand bath by means of a metal support, capable of holding four reactors at a time. The reactors were wrapped with aluminum foil to enhance the heat transfer and to protect the reactor from sand. The metal support was connected to a Dayton Permanent Magnet DC Motor. The speed of the motor can be controlled by means of a Penta Power multi drive speed controller. The reaction time was measured by means of a stop watch, which was started immediately after immersing the reactors into the sand bath. After reaching the desired reaction time, the reactors were removed from the sand bath and were immediately quenched in cold water.

After cooling the reactors in the cold water, the gas from each reactor was released into the Fourier Transform Infrared analyzer gas cell for analysis. After this, the reactor was opened and the liquid was removed from the reactor and stored in a sample bottle for later analysis. The reactor was washed with a known amount of acetone.

Analytical Methods

An HP5890 Gas Chromatograph (GC) and Nicolet Impact 400D Fourier Transform Infrared Analyzer (FTIR) were used for the liquid phase and gas phase analysis respectively.

An HP-17 Crosslinked 50% Phenyl Methyl Siloxane column and flame ionization detector was used for the liquid phase analysis. The GC oven was maintained initially at a temperature of 40°C for 6 minutes and was then heated to a temperature of 225°C at a rate of 5°C/minute. It was maintained at this temperature for 15 minutes. Helium was used as the carrier gas. A 1:20 split ratio was used, and the column flow rate was maintained at 5 mL/minute at 50°C. The auxiliary gas flow rate was fixed at 25 mL/minute. An internal standard, 1,2-dichlorobenzene, was used for the calibration and analysis.

The gas phase was analyzed on a Nicolet Impact 400D FTIR. A 120 mL gas cell was used for the analysis. The instrument was calibrated using standard gases of known concentration for carbon dioxide, carbon monoxide and nitrous oxide. The detection limit was 10 ppm for all the three components.

Results and Discussion

All the experiments were conducted at a pressure of 242 atm. The amount of catalyst was fixed at 2.5 mg. The effect of retention time, excess oxygen, external agitation, and catalyst particle size on the pyridine conversion were studied. The oxidation reaction of pyridine to ultimate products, carbon dioxide, nitrogen and water is given by

$$C_5H_5N + \frac{25}{4}O_2 \rightarrow 5CO_2 + \frac{1}{2}N_2 + \frac{5}{2}H_2O \tag{1}$$

The excess oxygen was defined as the percentage excess number of moles of oxygen used than the stoichiometric requirement. The retention time was varied between 5 and 20 minutes, the excess oxygen was varied between 100 and 500%, the external agitation was varied between 465 and 1860 rpm, and the particle size was varied between 0.15 mm and 1.5 mm. The various sizes of the catalyst was obtained by crushing the original catalyst. For all the experiments, the concentration of pyridine was fixed at 0.0075 mol/L.

Figure 1 shows the effect of retention time and excess oxygen on the supercritical water oxidation of pyridine in the absence of an added catalyst. In this figure, the lines represent the model predictions, and the markers represent the experimental values. The conversion increased with both the retention time and excess oxygen. The conversion increased from 41.9% to 78.5% as the retention time increased from 5 to 20 minutes at an excess oxygen of 300%. The major product observed was carbon dioxide, with the minor products being carbon monoxide and nitrous oxide. The yield of carbon dioxide (defined as the moles of carbon dioxide formed per mole of carbon present in the feed) varied between 2 and 0.25. The yield of carbon monoxide varied between 1 and 0.03, whereas the yield of nitrous oxide varied between 0.15 and 0.03. Additional minor products were tentatively identified by GC-MS including 2,4,6-trimethyl benzenamine, 3,4-dihydro 2H-1,4-benzoxazine diethyl phthalate, and 3-butylpyridine 1-oxide. The yield of all minor products was less than 0.01 based on the peak areas observed on the GC. The carbon balance varied between 1 and 0.6, while the nitrogen balance was between 0.8 and 0.3. The lowest mass balances were obtained at 100% excess oxygen and 5 min retention time, and the mass balance improved with an increase in the excess oxygen and the retention time. Because these experiments were run with air, nitrogen could not be confirmed in these experiments. This might account for the low nitrogen balance.

The effect of retention time on the supercritical water oxidation of pyridine in the presence of MnO_2/CeO_2 catalyst is shown in Figure 2. This experiment was conducted at a temperature of 400°C with the smallest catalyst particles of 0.15 mm and an external agitation rate of 430 rpm. Again, the lines represent the model predictions, and the markers represent the experimental values. The multiple points represent the data obtained from different reactors at identical conditions. The conversion increased from 69% to 95% as the retention time was increased from 5 to 20 minutes at an excess oxygen of 100%. Figure 2 also shows the effect of excess oxygen. Increasing the excess oxygen led to an increase in the conversion of pyridine. At all reaction conditions carbon dioxide was the major product. The other gas phase products that were observed were carbon monoxide and nitrous oxide. The yield of carbon dioxide varied between 3 and 0.5. The yield of carbon monoxide varied between 0.2 and 0.03, whereas the yield of nitrous oxide varied between 0.4 and 0.05. The other minor liquid phase products identified by GC-MS include diethyl phthalate, 3-butylpyridine 1-oxide, hexanedioic acid dioctyl ester. The yield of these products was always less than that observed for the homogeneous case. The carbon balance was between 1 and 0.6, whereas the nitrogen balance was between 0.8 and 0.5.

The conversion data was fit to an empirical power law rate expression of the form

$$r = -\frac{1}{W}\frac{dN_{Pyr}}{dt} = -\frac{V}{W}\frac{dC_{Pyr}}{dt} = kC_{Pyr}^n C_{O_2}^m \qquad (2)$$

where r is the rate of reaction expressed as moles of pyridine reacted per gram of catalyst per second, k is the rate constant, $(mol/cm^3)^{1-m-n}\ cm^3\ g^{-1}\ s^{-1}$, C_{pyr} is the concentration of pyridine, (mol/cm^3) C_{O2} is the initial concentration of oxygen, and W is the weight of catalyst in grams. In the case of homogeneous reactions, the catalyst weight was replaced with the volume of reactor and the rate of the reaction was expressed as the moles of pyridine reacted per cubic centimeter per second. Since oxygen was present in excess all the time, the oxygen concentration was assumed to be constant at the initial value. An integral analysis gave the values of k, n, and m. The rate of reaction of pyridine oxidation on MnO_2/CeO_2 catalyst in SCW at 400°C was found as

$$r = 5.47 \times 10^4\ C_{Pyr} C_{O_2}^{1.04} \qquad (3)$$

where the unit of the rate constant was $(mol/cm^3)^{-1.04}\ cm^3\ g\ cat^{-1}\ s^{-1}$. The rate of reaction of SCWO of pyridine was found as

$$r = 1.87 \times 10^{-2}\ C_{Pyr} C_{O_2}^{0.3} \qquad (4)$$

where the unit of the rate constant was $(mol/cm^3)^{-0.3}\ s^{-1}$. As shown in Figures 1 and 2, a good agreement was obtained from the model predictions.

A direct comparison between the homogeneous reaction and the catalytic reaction at 300% excess oxygen is shown in Figure 3. For both cases, an increase in the retention

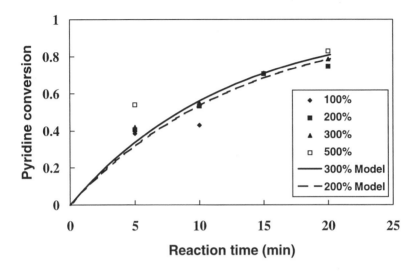

Figure 1: The effect of retention time and excess oxygen on supercritical water oxidation of pyridine without the addition of heterogeneous catalyst.

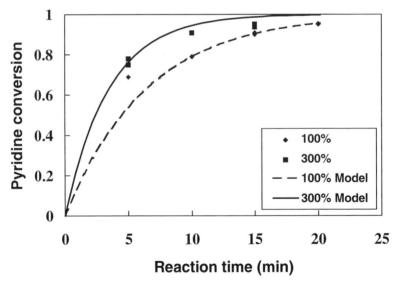

Figure 2: The effect of retention time and excess oxygen on supercritical water oxidation of pyridine in the presence of MnO_2/CeO_2 catalyst.

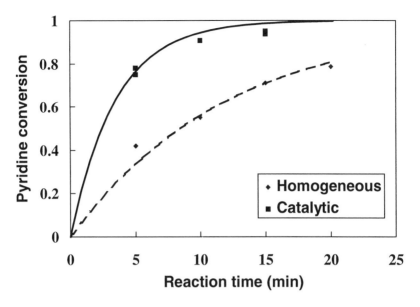

Figure 3: Direct comparison between catalytic and homogeneous oxidation of pyridine at 300% excess oxygen, lines indicate model predictions.

time led to an increase in the pyridine conversion. At all oxygen concentrations and retention times higher conversions were obtained for the catalytic case. For example, at 400°C, 300% excess oxygen, and a retention time of 5 min the conversion increased from 40% for the homogeneous case to 78% in the presence of MnO_2/CeO_2 catalyst. A maximum of 95% conversion was obtained after 15 minutes of catalytic reaction, whereas the maximum conversion that was obtained for the homogeneous case was 80% after 20 minutes of reaction.

As mentioned previously, Crain *et al.*, (7) reported a maximum conversion of 68% at a temperature of 527°C and a residence time of 3.84s. They reported no conversion of pyridine below 425°C, and obtained a conversion of 3% at 426°C and a residence time of 10.5s. In these experiments, the oxygen to pyridine molar ratio was varied between 1 and 2. They also reported that the rate of reaction of pyridine in SCW to follow

$$r = 1.26 \times 10^{13 \pm 1.65} \exp(-209.5 \pm 22.4 \, / \, RT) C_{Pyr} C_{O_2}^{0.2} \tag{5}$$

with the units of the rate constant and activation energy being $(mol/L)^{-0.2}$ s^{-1} and kJ/mol respectively. In our experiments the oxygen to pyridine molar ratio was varied between 25 and 50 for both homogeneous and catalytic reactions. A more general rate expression of the form

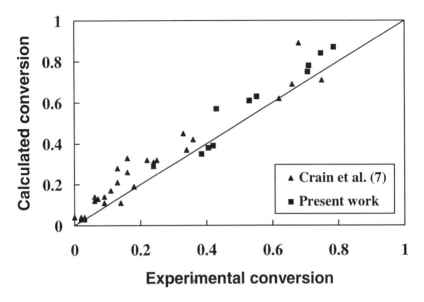

Figure 4: Parity plot, comparing the current experimental data and that obtained from the literature with the model predictions for the conversion of pyridine during homogeneous SCWO.

$$r = 5.17 * 10^{17} \exp(-250.3 / RT) C_{Pyr} C_{O_2}^{0.28} \tag{6}$$

was found to better describe the data from the present work as well as from the work of Crain et al., (7). Here the units of the rate constant and the activation energy are $(\text{mol/cm}^3)^{-0.28}$ s^{-1} and kJ/mol respectively. A comparison between the experimental and calculated conversion is shown in Figure 4. A good agreement was obtained.

Mass Transfer effects

In general the rate of the catalytic reaction can be limited by the transport of the reactants from the bulk phase to the catalyst surface (external mass transfer), or the transport of the reactants inside the catalyst pores (internal mass transfer). In the absence of these mass transfer limitations, intrinsic kinetics can be obtained. In general, there are two different ways to determine the presence of mass transfer gradients (9). The first method is to conduct an experiment. This can be done by studying the effect of catalyst particle size for internal mass transfer limitations or by varying the fluid velocity (or rate of agitation) for external limitations. The second method is by calculating the mass transfer coefficient and diffusivity. This method is limited by the accuracy with which the physical properties can be predicted, by the knowledge of the basic kinetics, and by the complexity of the reaction.

External Mass transfer effects. In this regime, the reaction rate is controlled by the mass transfer of the reactant from the bulk fluid to the surface of the catalyst. The concentration of the reactant on the catalyst surface approaches zero, and the observed rate is the mass transfer rate, which is smaller than the intrinsic reaction rate. The apparent activation energy is about 4 to 12 kJ/mol for gases, 10 to 20 kJ/mol in liquid hydrocarbons and 8 to 10 kJ/mol in aqueous systems. In this regime all reactions appear to be first order regardless of their intrinsic kinetics, since mass-transfer is a first order process (9).

In our batch reactor system, experiments were conducted to study the effect of the external agitation on the conversion of pyridine. These experiments can be used determine if external mass transfer limitations exist. In our reactor system, the extent of internal mixing is proportional to the external agitation. Hence the internal mixing of the fluid and the solid can be increased by increasing the rate of external agitation. Under external mass transfer limitation, an increase in the internal mixing will lead to an increase in the conversion. In the absence of any effect of external mass transfer, the rate or conversion should be independent of external agitation.

The effect of external agitation on the conversion of pyridine was studied at 380°C and a constant retention time of 15 min. The results are shown in Figure 5. As the external agitation was increased from 430 to 1380 rpm, the conversion increased from 61% to 83% at an excess oxygen of 100%. In this experiment the particle size was less than 0.15 mm. A similar trend was observed at 500% excess oxygen and is also indicated

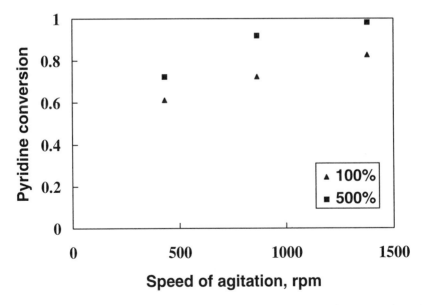

Figure 5: The effect of external agitation of the reactor on the conversion of pyridine during catalytic SCWO at 380°C.

in Figure 5. A similar effect was seen on the yield of carbon dioxide. The results qualitatively indicate that the experiments were conducted in a regime where external mass transfer was important.

Internal mass transfer limitations. In this regime the reaction rate is controlled by the transport of the reactants from the catalyst surface to the active sites within the catalyst. In this case, the diffusion process is not independent of the reaction rate.

The presence of internal mass transfer effects can be determined by studying the effect of particle size on the rate of reaction or reactant conversion. Under the influence of internal mass transfer, the observed rate of reaction is inversely proportional to particle size. In the absence of the internal mass transfer limitation, the observed rate will be independent of particle size. For intermediate degrees of internal mass transfer limitation, the observed rate will decrease with increasing particle size, although not by the inverse relationship observed with strict mass transfer control.

The effect of particle size on the supercritical water oxidation of pyridine was studied at 380°C, at 100% and 500% excess oxygen, at an external agitation speed of 863 rpm, and for a retention time of 15 min. The effect of particle size on the conversion is shown in Figure 6. For the smallest catalyst particles (less than about 0.4 mm), the conversion was nearly independent of particle size. However, for larger particle sizes, pyridine conversion decreased as the particle size increased. This suggests that for the smallest catalyst particles, the rate of intraparticle mass transfer was substantially greater than the rate of reaction. For larger particles, on the other hand, the reaction was influenced by intraparticle mass transfer limitations.

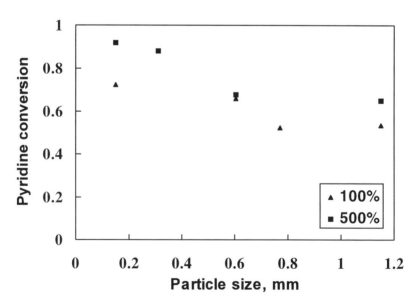

Figure 6: The effect of catalyst particle on the conversion of pyridine during catalytic SCWO at 380°C.

Conclusions

The effect of a catalyst on the supercritical water oxidation of pyridine was studied. The addition of MnO_2/CeO_2 catalyst increased the pyridine conversion, the yield to ultimate products namely, carbon dioxide, water, nitrogen and nitrous oxide. Several partial oxidation or condensation products were detected; their yield was decreased by the addition of the catalyst. The data was described by an empirical power law rate expression. Both external and internal mass transfer limitations were found to be present during this reaction.

Acknowledgments

This research has been supported by the National Science Foundation under grant number CTS-9317618.

Literature Cited

1. Barner, H.E., Huang, C.Y., Johnson, T., Jacobs, G., Martch, M.A., "Supercritical water oxidation: An emerging technology", *Journal of Hazardous Materials*, **31**, 1-17, 1992.

2. Quist, A.S., "The Ionization Constant of Water to 800°C and 4000 bars", *Journal of Physical Chemistry*, **74**, *18*, 3396, 1970.

3. Marshall, W.L., Franck, E.U., "Ion Product of Water Substance , 0-1000°C, 1-10,000 bars, new international formula and its background", *Journal of Physical and Chemical Reference Data*, **10**, *21*, 295-304, 1981.

4. Franck, E.U., "Water and Aqueous Solutions at High Pressures and Temperatures", *Pure and applied Chemistry*, **24**,*13*,1970.

5. Japas, M.L., Franck, E.U., "High Pressure phase equilibria and PVT-data of the water-nitrogen system to 673 K and 250 MPa", *Ber. Buns. Phys. Chem.*, **89**, *7*, 793-800, 1985.

6. Dell'Orco, P.C., Gloyna, E.F., Buelow, S., "Oxidation Processes in the Separation of Solids from Supercritical Water", in *Supercritical Fluid Engineering Science: Fundamentals and Applications*, Edited by Kiran, E., Brennecke, J.F., ACS Symposium Series 514, 1992.

7. Crain, N., Tebbal, S., Li, L., Gloyna, E.F., "Kinetics and Reaction Pathways of Pyridine Oxidation in Supercritical Water", *Industrial and Engineering Chemistry Research*, **32**, 2259-2268, 1993.

8. Wightman, T. J., *"Studies in Supercritical Wet Air Oxidation"*, Master of Science Thesis, University of California, Berkeley, 1981.

9. Satterfield, C.N., *"Heterogeneous Catalysis in Industrial Practice"*, Second Edition, McGraw Hill, Inc., 1991.

Chapter 17

Degradation in Supercritical Water Oxidation Systems

D. B. Mitton, E.-H. Han, S.-H. Zhang, K. E. Hautanen, and R. M. Latanision

H. H. Uhlig Corrosion Laboratory, Department of Materials Science
and Engineering, Massachusetts Institute of Technology, 77 Massachusetts
Avenue, Cambridge, MA 02139–4307

Supercritical water oxidation (SCWO) can effectively destroy various
civilian and military wastes; however, the system will generally need
to withstand a corrosive environment. To improve our understanding
of degradation within such systems, exposure testing is being carried
out in conjunction with analysis of failed components. Various alloys
have been exposed to environments ranging from deionized water to
highly chlorinated organic compounds and to temperatures as high as
600°C. Although, not surprisingly, high corrosion rates are
encountered for the chlorinated feed streams, even deionized water can
be aggressive at these conditions. In chlorinated feed streams,
Hastelloy C-276 exhibits premature failure at subcritical conditions as
a result of dealloying and cracking. At supercritical conditions
(600°C) in chlorinated environments, both high-nickel alloys and
stainless steel exhibited significant corrosion. Analysis suggests that it
may be possible to alter feed characteristics to reduce degradation to
an acceptable level. This may permit the use of less costly materials
for construction

At a time when public opposition to landfills and incineration is increasing, the cleanup of military and civilian hazardous wastes is gaining national importance (1).
Supercritical water oxidation (SCWO) is one promising technology applicable to
many organic wastes (2-4), including dilute aqueous solutions, which are difficult to
treat by conventional methods. SCWO capitalizes on the properties of water
characteristic of supercritical conditions, (374°C and 221 atm for pure water), where
water is a fluid possessing properties between those of a liquid and a gas. As the
solvation properties of supercritical water resemble those of a low polarity organic (2)
hydrocarbons exhibit a significant increase in solubility concurrent with a reduced
solubility of inorganic salts.
  Unfortunately, corrosion of the materials of fabrication is a significant concern in
the context of the development of scaled-up systems for supercritical water oxidation
(5-7) and may ultimately be the deciding factor in the commercial application of this
technology. Although high nickel alloys may be employed for severe service (8),
they are apparently unable to withstand certain conditions associated with SCWO (5,
9-14) as they tend to exhibit both significant weight loss and localized effects

including stress corrosion cracking (SCC) and dealloying in chlorinated environments. In addition, selective dissolution of various alloying components from the high nickel alloys has been observed in SCWO systems. The loss of Cr and Mo for Inconel-625 (*15*) has been reported while, for Hastelloy C-276, the loss of either Cr, Mo and W (*15*) or Ni, Fe and Mo (*16*) is observed.

Exposure studies were previously carried out in deionized water during commissioning of the corrosion test facility at MIT to provide a base-line assessment of the materials of construction. The results (*6, 13*) of a preliminary short term exposure (24 - 96 hours) at 300, 400 and 500°C indicate varying degrees of film formation and localized corrosion for the three alloys tested (Inconel-625, Hastelloy C-276 and stainless steel type 316). The effect of exposure time (up to 240 hours), in deionized water at 500°C (*12*) was interpreted to indicate the formation of a protective oxide for both Inconel-625 and Hastelloy C-276 and a non-protective film for the stainless steel. During tests with methylene chloride (CH_2Cl_2), axial through-wall ruptures have occurred in a number of Hastelloy C-276 preheater tubes of a tubular plug-flow reactor (PFR) system employed for kinetic studies at MIT. Failures occurred in a region of the preheater which was at a high subcritical temperature and preliminary analysis revealed a dealloyed layer in which Ni (the major component of Hastelloy C-276) was lost leaving primarily a chromium oxide (*10, 16*).

Results and Discussion.

Exposure Studies. A series of experiments conducted in deionized water revealed the possibility of pit initiation in high nickel alloys such as Inconel-625 and Hastelloy C-276 (*12*). Figure 1 reveals a recent scanning confocal laser micrograph of a pit on Inconel-625 after 10 days at 500°C in deionized water. The surface profile reveals that the depth of the pit is 2.64 μm and its width is slightly more than 6 μm. Similar localized phenomena were observed on Hastelloy C-276; however, the pits tended to be covered by a reaction product cap (*12*). Although these pits are not large and exhibit a penetration rate of approximately 4 mils per year (mpy), the development of localized phenomena in such an innocuous environment is of interest. This suggests the possibility of the development of localized corrosion during system down-time (if the system is flushed with water) or during passivation treatments which employ water. Such discrete regions are potential stress concentration sites and may result in problems during subsequent runs with aggressive feed solutions.

Recently various alloys have been exposed to a highly chlorinated organic feed stream at 600°C. Figure 2 presents the corrosion rate based on weight loss for an exposure time of 66.2 hours. The feed stream contained approximately 3000 mg/kg chloride and 6 wt% O_2. Based on weight loss data, C-276 and Inconel 625 exhibit similar corrosion rates of about 700 mpy while stainless steel type 316-L reveals a rate in excess of 2000 mpy. Figures 3 (a-c) present the cross section of U-bend samples after exposure. The weight loss data is corroborated by the stainless steel sample (Figure 3(a)), which has lost a significant portion of the cross section. Both Inconel-625 (Figure 3(b)) and Hastelloy C-276 (Figure 3(c)) exhibit a lower loss of cross section than the stainless steel, which again agrees with the weight loss data. The loss of cross section is, however, less on the Inconel-625 sample than on the Hastelloy C-276 coupon. This finding is in agreement with previously reported results indicating that Inconel-625 is more corrosion resistant than C-276 at temperatures above 350°C (*5*). Microscopic examination of the U-bend samples revealed no indication of stress corrosion cracking (SCC) for either of the nickel alloys; however, cracks were observed for the stainless steel sample. At high subcritical (*10*) and low supercritical (*5*) temperatures, Hastelloy may exhibit cracking. When exposed at 580°C to a feed stock of methylene chloride and isopropyl alcohol (neutralized with NaOH), Inconel 625 exhibited SCC; however,

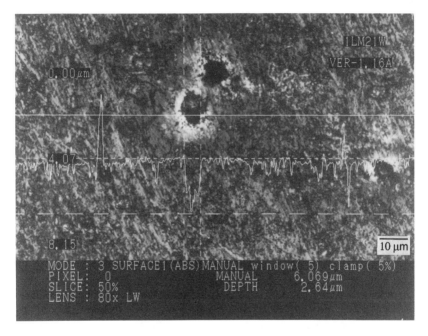

Figure 1. Scanning confocal laser micrograph of a pit on Inconel-625 after 10 days at 500°C in deionized water.

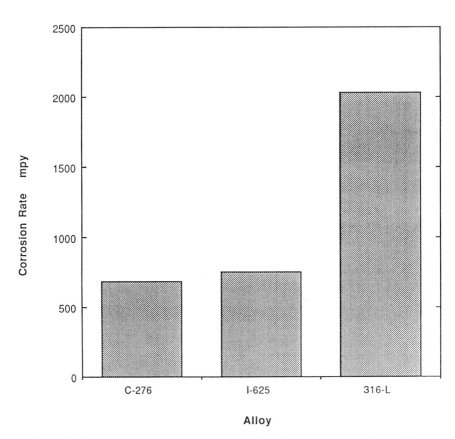

Figure 2. The corrosion rate based on weight loss for an exposure time of 66.2 hours. The feed stream contained approximately 3000 mg/kg chloride and 6 wt% O_2.

Figure 3. The cross section of (a) 316-L stainless steel, (b) Inconel-625 and (c) Hastelloy C-276 U-bend samples after exposure for 66.2 hours to a feed stream contained approximately 3000 mg/kg chloride and 6 wt% O_2.

cracking was not observed for an Inconel-625 reactor exposed to a variety of feed streams even after 3000 hours of service (5).

Failure Analysis. Premature failure of two preheater tubes from an experimental system (Figure 4) employed to study the kinetics of destruction in SCWO at MIT have resulted during experiments which included ambient methylene chloride (CH_2Cl_2) feed concentrations ranging from 0.017-0.04 Mol/l. The total time of exposure to the Cl⁻ containing feed solution prior to failure was 104 hours for the first and 45 hours for the second preheater. The Hastelloy C-276 preheaters were 250 - 300 cm long had an outside diameter (OD) of 1.6 mm and a wall thickness of 0.254 mm.

Although, ultimately, failure took the form of an axial intergranular crack (*10*), elemental depletion (Ni, Mo, Fe) of the corroded layer and elevated Ni concentrations in the effluent in conjunction with no observable change in the dimensions of the corroded and non-corroded regions of the same tube confirmed that dealloying was also occurring for this alloy at these conditions (*16*). Although different conditions (temperature, feed type and composition) were employed during the various experiments, failures were restricted to a 20 cm length of tube near the inlet end of the preheater (Figure 4). Experimental conditions were designed for temperatures within the reactor section of the system to be between about 450 and 600°C; however, in the preheater, temperature varies between ambient and the test condition. Although at supercritical temperatures, the outlet of each preheater suffered minimal degradation (*10*) amounting to approximately 20 mpy. The lack of any appreciable corrosion at the exit to the preheater, which experienced significant HCl concentrations at supercritical conditions, in conjunction with the proximity of the failures (within 20 cm of each other) suggested that failures were occurring within a region of the tube which was at subcritical temperatures. This was confirmed by heat transfer calculations indicating that the failure site was at a high but subcritical temperature (*16*).

More recently a preliminary investigation of the cool-down heat exchanger from this system (Figure 4) has provided some insight into corrosion within these facilities. Although the supercritical exit to the preheater suffered minimal attack, significant corrosion was subsequently observed near the inlet to the cool-down heat exchanger. The form of attack in this region (Figure 5) is very similar to that seen within the subcritical region of the preheater (*10, 16*). Again intergranular dealloying is apparent and grain boundaries are obvious in the dealloyed layer. This is an important observation suggesting that there is a relatively restricted temperature region, apparently corresponding to the temperature transition between subcritical and supercritical, within which corrosion is most severe. This situation is presented in Figure 6 which schematically reveals the rate of corrosion as a function of temperature within the preheater and cool-down heat exchanger. In the preheater the corrosion rate echoes the temperature as it increases into the high subcritical range; however, a further increase in temperature to supercritical, results in a dramatic decrease in the rate of degradation. The type and extent of corrosion within the reactor is not known as this unit has never been removed from service. It seems likely, however, that degradation relative to the preheater, is insignificant as this unit has exhibited no problems while, during the same time period, a number of preheaters have failed. In addition, the exit to the preheater and inlet to the cool-down heat exchanger, which are at supercritical conditions, both reveal low rates of corrosion. This suggests a similar condition within the supercritical reactor , which is located between these two points. A short distance into the cool down heat exchanger the corrosion rate increases again at a point which is assumed to be at a high subcritical value. Finally, beyond this maximum the corrosion rate decreases as a function of decreasing temperature. From the corrosion standpoint, this suggests the possibility

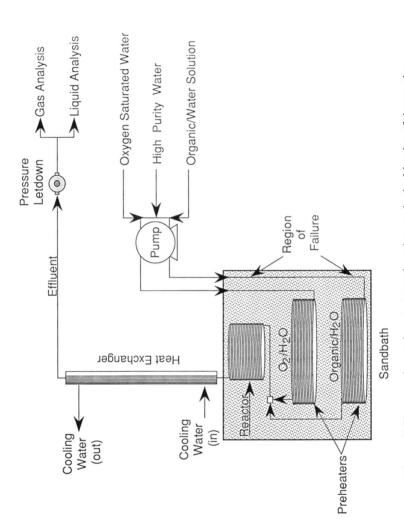

Figure 4. The experimental system employed to study the kinetics of destruction in SCWO at MIT. Reproduced with permission from reference 16. Copyright 1996 The Electrochemical Society, Inc.

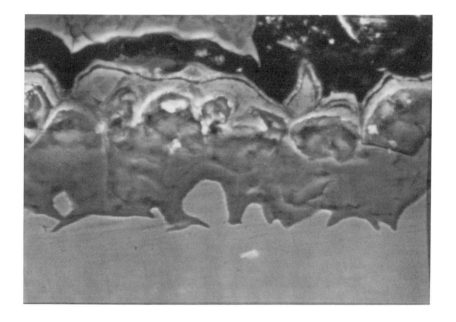

20 μm

Figure 5. Cross-section of a Hastelloy C-276 cool-down heat exchanger used in the system presented in the previous figure.

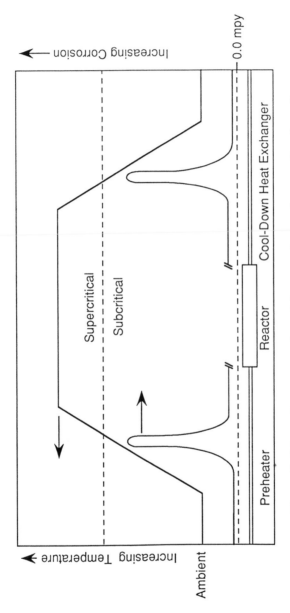

Figure 6. Schematic representation of the rate of corrosion as a function of temperature within the preheater and cool-down heat exchanger sections of the system presented in Figure 4.

of a design which could incorporate an easily replaced inexpensive section located at the two corrosion maxima.

E-pH Diagrams. One of the main tools employed during an initial assessment of the possible behavior of a metal is the E-pH (Pourbaix) diagram. These maps graphically present the correlation between the solution oxidizing strength (potential) and acidity (pH) for the various stable phases possible for a system. Normally these diagrams are relevant for ambient conditions; however, data are available for some systems at higher temperatures (*17, 18*). These diagrams are constructed by using thermodynamic data available in the literature and the Nernst equation. It is, thus, possible to delineate regions in which specific species are most stable. This can be simplified further into regions which thermodynamically favor (i) immunity (no corrosion), (ii) passivity (reduced corrosion as a result of a protective film) and (iii) corrosion. By employing these diagrams the thermodynamically favored condition can be observed for a given pH and potential. It is possible to incorporate a number of such diagrams for various temperatures into one E-pH-T diagram, which can then be used to provide information on the trends in thermodynamic stability as a function of temperature. This is particularly valuable in the case of supercritical water oxidation as although the main reactor is at a higher temperature, some system components will experience lower temperatures ranging between ambient and the temperature of operation. As previously discussed, these systems can exhibit severe corrosion, which may be intensified within a restricted temperature zone.

Figures 7 (a) and (b) present the chromium and nickel E-pH-T diagrams respectively. In all cases a molar activity of 1×10^{-6} was selected for calculation of the equilibrium line between soluble and insoluble species. Calculations were accomplished using the $\Delta G°$ values available for the individual temperatures in the literature (*17*) or through private communication (*19*). Figure 7(c) presents the region within which both chromium and nickel are stable over the temperature range 100°C to 300°C. If this region could be maintained during supercritical water oxidation it would minimize the likelihood of dealloying. This could have a significant impact on the use of alloys such as Hastelloy C-276 and may obviate the need for exotic liners (*20*).

Although they do provide valuable information, the limitations of E-pH diagrams must be recognized; in general, they refer to pure, defect free, unstressed metals in pure water. While they indicate reactions which are thermodynamically possible, they do not provide any information on the reaction rate. They must, therefore, be utilized with great caution in predicting potential corrosion behavior.

Conclusions.

Examination of U-bend samples (316-L stainless steel, Inconel-625 and Hastelloy C-276) revealed no indication of stress corrosion cracking for the nickel alloys after exposure at 600°C for 66.2 hours to a feed stream contained approximately 3000 mg/kg chloride and 6 wt% O_2. The stainless steel sample, however, exhibited cracking. Weight loss data from the same experiment indicate a very high corrosion rate (2000 mpy) for the stainless steel. Although the two high nickel alloys exhibit lower corrosion rates (700 mpy) than the stainless steel, the rate is still very high.

Two corrosion maxima were observed during a failure analysis of Hastelloy C-276 tubing employed in a PFR system at MIT. These maxima apparently correspond to the temperature transition between subcritical and supercritical, within which corrosion is most severe. Potentially this result indicates the possibility of a system design which could incorporate an easily replaced inexpensive section located at the two corrosion maxima.

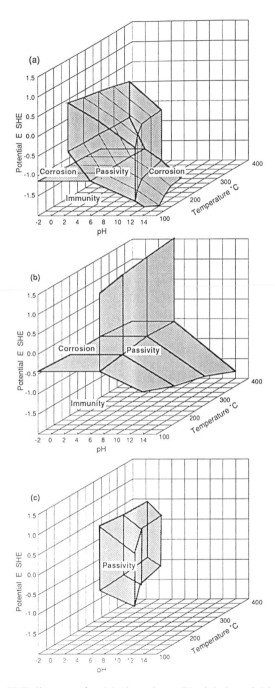

Figure 7. E-pH-T diagrams for (a) chromium, (b) nickel, and (c) the region of stability for both chromium and nickel.

E-pH-T diagrams were developed which reveal the region of stability for Cr and Ni over the temperature range between 100 and 300°C. The ability to maintain this region within SCWO systems could obviate the need for exotic liners and possibly permit the use of high nickel alloys as materials of fabrication.

Acknowledgments.

The support of the Army Research Office for this project is much appreciated. The assistance of Glenn T. Hong of MODAR and Ron E. Mizia of INEL is gratefully acknowledged. K.E. Hautanen is supported by an AASERT award. E-H Han is supported, in part, by a Distinguished Senior Visiting Fellowship from the Chinese Academy of Science.

Literature Cited.

1. Swallow, K.C. and Ham, D., *The Nucleus*, 1993, 11, p.11.
2. Tester, J.W.; Holgate, H.R.; Armellini, F.J.; Webley, P.A.; Killilea, W.R.; Hong, G T. and Barner, H.E., in *Emerging Technologies for Waste Management III*, ACS Symposium Series, 518, ACS , Washington, DC 1993 p.35.
3. Modell, M., *Standard Handbook of Hazardous Waste Treatment and Disposal* , McGraw Hill, New York NY 1989 p. 8.153.
4. Franck, E.U. *High Temperature, High Pressure Electrochemistry in Aqueous Solutions*, NACE, Houston TX 1976 p. 109.
5. Latanision, R. M. and Shaw, R.W., Co-Chairs, "Corrosion in Supercritical Water Oxidation System - Workshop Summary"; 1993, Massachusetts Institute of Technology Energy Laboratory, MIT-EL 93-006.
6. Mitton, D.B.; Orzalli, J.C.; and Latanision, R.M., in *Proceedings of the Third International Symposium on Supercritical Fluids* , ISASF, Nancy France, 1994 Vol.3, p.43.
7. Thomas, A.J. and Gloyna, E.F. "Corrosion Behavior of High Grade Alloys in the Supercritical Water Oxidation of Sludges," 1991, University of Texas at Austin, Technical Report CRWR 229.
8. Asphahani, A., *Metals Handbook 9th Edition;* ASM International: Metals Park, OH, 1987, Vol. 13 p 641–655.
9. Norby, Brad C. "Supercritical Water Oxidation Benchscale Testing Metallurgical Analysis Report," 1993, Idaho National Engineering Laboratory Report, EGG-WTD-10675
10. Latanision, R.M., *Corrosion*, 1995, 51, p. 270.
11. Mitton, D.B.; Orzalli, J.C.; and Latanision, R.M., in *Innovations in Supercritical Fluids: Science and Technology*, ACS Symposium Series, 608, ACS, Washington, DC 1995 p. 327.
12. Mitton, D.B.; Orzalli, J.C.; and Latanision, R.M., in *Physical Chemistry ofAqueous Systems-Meeting the Needs of Industry*, Proc. 12th ICPWS, Begell House New York, NY 1995, p.638.
13. Orzalli, J.C., "Preliminary Corrosion Studies of Candidate Materials for Supercritical Water Oxidation Reactor Systems" 1994, Master's Thesis, Massachusetts Institute of Technology.
14. Kane, R.D. and Cuellar, D., "Literature and Experience Survey on Supercritical Water Corrosion," 1994, CLI International Report, No L941079K.
15. Bramlette, T.T.; Mills, B.E.; Hencken, K.R.; Brynildson, M.E.; Johnston, S.C.; Hruby, J.M.; Feemster, H.C.; Odegard, B.C.; and ucdd, M., "Destruction of DOE/DP Surrogate Wastes with Supercritical Water Oxidation Technology", 1990, Sandia National Laboratories, Livermore, CA, Sand 90-8229.
16. Mitton, D.B.; Marrone, P.A.; and Latanision, R.M., *J. Electrochem Soc.* 1996,Vol. 143 pp. L59.

17. Lee, J.B., *Corrosion* , **1981**, 37, p. 467.
18. Kriksunov, L. B. and Macdonald, D.D. "Development of Pourbaix Diagrams for Metals in Supercritical Aqueous Media," presented at The First International Workshop on Supercritical Water Oxidation, Amelia Island Plantation, Jacksonville Florida, February 6-9 **1995**.
19. Kriksunov, L. B., private communication **1995**.
20. Hazlebeck, D.A.; Downey, K.W.; Jensen, D.D.; and Spritzer, M.H., in *Physical Chemistry ofAqueous Systems-Meeting the Needs of Industry*, Proc. 12th ICPWS, Begell House New York, NY **1995**, p.632.3.

Chapter 18

Supercritical Carbon Dioxide Extraction of Solvent from Micromachined Structures

Edward M. Russick[1], Carol L. J. Adkins[1], and Christopher W. Dyck[2]

[1]Organic Material Processing Department and [2]Microelectronics Technologies Department, Sandia National Laboratories, P.O. Box 5800, Albuquerque, NM 87185

We have demonstrated that supercritical carbon dioxide extraction can be used for solvent removal to successfully release compliant surface micromachined structures on silicon wafers developed at Sandia National Laboratories. Structures that have been successfully extracted and released include single gear microengines, bridge and cantilever beams, pressure transducers, and experimental comb drive actuators. Since the supercritical fluid has negligible surface tension, it has virtually unabated access to solvent residing in capillary-like spaces as narrow as 1-3 μm under the micromachined features. While conventional drying techniques have been plagued with the collapse and sticking of micromachined structures due to surface tension effects, supercritical carbon dioxide has been shown to reproducibly dry components and test structures, including bridge and cantilever beams approaching 1000 μm in length, without collapsing. The equipment and the extraction process are described, and photographs of supercritically dried test structures and components are presented.

With the trend towards miniaturization of microprocessors in the electronics industry, it seems appropriate that the development of miniature electromechanical components should be of interest as well. In fact, surface micromachining of polysilicon films deposited on silicon wafers is an emerging technology in the fabrication of microactuators and microsensors (1). These miniaturized components include microengines, microlever actuators, accelerometers, and pressure sensors which have potential uses in a variety of applications for mechanical and electrical devices both in industry and in government research (e.g., weapons design). Only surface micromachined structures are discussed, which involve the deposition and patterning of films above the surface of the silicon substrate (2). Bulk micromachining, which involves the removal of the substrate to define features (3) will not be discussed in this work.

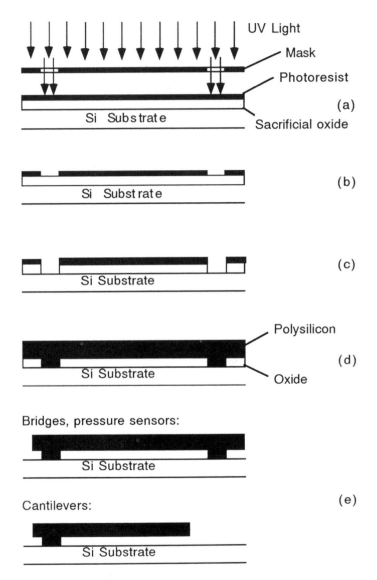

Figure 1. The basic micromachine etching process. A sacrificial oxide layer and photoresist are deposited on the silicon wafer, then the photoresist is exposed to UV light through a mask (a). The oxide layer is exposed where the photoresist was removed (b). Plasma etching removes the exposed oxide. (c). The photoresist is removed, then polysilicon is deposited (d). HF etch leaves the polysilicon in the desired configuration (e).

Surface machined microstructures are formed using a combination of masking, dry plasma etching of polysilicon film deposited on the wafer, and wet etching done in a liquid-phase acid solution such as hydrofluoric acid (HF). Figure 1 illustrates the basic surface micromachining process. The final HF etch is followed by a water rinse. In some cases the HF etch is followed by an ammonium fluoride (NH_4F) treatment (*4*), to be discussed later. This treatment may increase the long term reliability of the micromachined parts. After etching and rinsing, the part are dried to yield the released micromachined sample. It is to the drying process that we have applied supercritical fluid extraction technology.

A scanning electron micrograph of a microengine is shown in Figure 2. Microengines are multi-layer structures that are fabricated by repeating surface micromachining process in a series of successive layers (*1*). It can be seen that the many intricate features of the microengine have been formed on the silicon wafer in an area of only about 4 square millimeters. The function of the microengine is to drive a microgear, thus performing mechanical work at a microscopic scale. The microengine is driven by an alternating electrical bias applied to comb drive actuators whose motion is translated to the microgear. The microgear could serve as a spinning portion of a gyroscope or be used to control optical shutters and mirrors. Microlever actuators also are designed to perform work on a very small scale, but with a different mechanical design. With an arrangement of lever arms and fulcrums, the movement of the drive mechanism is increased such that slight motion at the drive is greatly increased at the working part. A micromachined accelerometer is essentially a suspended silicon plate, fixed to sidewalls by silicon springs, which is placed under an electrical bias. During acceleration, the plate will attempt to deflect, changing the bias required between the plate and the ground polysilicon plane to maintain the plate's equilibrium position. Acceleration is detected by the change in the bias. Micromachined pressure sensors consist of vacuum sealed cavities under diaphrams that deflects according to an applied pressure. Pressure measurement is accomplished with piezoresistors on the surface of the diaphram. Micrographs of these microstructures will be presented later in the results section.

Silicon is a very practical micromechanical material in that it is capable of a great amount of flexibility before fracturing. However, the compliant nature of the silicon makes it susceptible to fabrication problems. A significant problem in the fabrication of the micromachined components is sticking of released structures to the substrate after they are dried using conventional air drying techniques. The sticking, combined with static friction which these parts experience has been termed stiction, a phenomenon commonly seen in magnetic storage media (*5*).

A number of phenomena may potentially cause microdynamic stiction of suspended microstructures, several of which will be identified here. Electrostatic forces due to electrostatic charging may cause sticking. These forces can be generated on the wafer due to etching, rinsing and drying (*6*). This is a non-equilibrium condition which usually dissipates over time or with contact between conducting surfaces. Second, a smooth surface finish may cause stiction. Smooth surfaces are more likely to stick, while surface roughness effectively increases the nominal separation between micromachined surfaces. Slight roughness on adjacent surfaces can reduce adhesion forces by several orders of magnitude (*7*). The surfaces of the polysilicon microstructures usually have a microscopically textured surface after etching, possibly due to the orientation of the grain structure of the polysilicon as it was deposited and etched. Physical alteration of contacting surfaces through dry plasma etching has also been reported (*8*). NH_4F etching, mentioned previously, has been demonstrated to slightly roughen the solid surfaces of the silicon surfaces. This may have a positive effect by inhibiting surface stiction and increasing the long term reliability of the microstructure. Third, a phenomenon

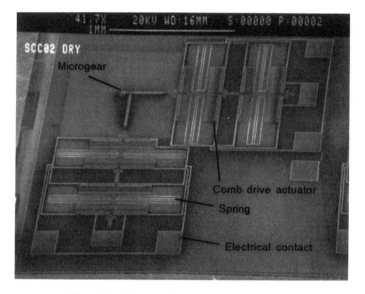

Figure 2. Microengine, magnified 41.7 times.

called solid bridging (*6*) occurs when non-volatile impurities present in the drying liquid are deposited on the surfaces of the microstructures. The impurities in narrow gaps formed by the suspended microstructures essentially bridge the gaps, causing the structures to stick. Obviously, avoiding impurities in the rinse liquid would help to minimize solid bridging.

Perhaps the most troublesome cause of surface stiction is liquid bridging (*6*). Liquid bridging is due to the surface tension effects of trapped capillary liquids upon drying. The liquid, usually water, used to rinse the microstructures is trapped in the narrow gaps between the silicon wafer and the suspended structures. Interfacial forces generated when the trapped capillary fluid dries can cause the microstructures to collapse and stick. Figure 3 shows a comparison of a stuck pressure sensor with one that has been successfully released. The meniscus force between two flat, polished surfaces with a liquid bridge is given by (*5*):

$$F_m = \frac{\gamma A}{h} (\cos\theta_1 + \cos\theta_2)$$

where θ_1 and θ_2 are the contact angles of the liquid with the two solid surfaces, A is the shared area of the parallel surfaces, assuming the gap between them is flooded with capillary liquid, h is the average thickness of the liquid bridge, and γ is surface tension, 73 dynes/cm for water. Obviously, reduction or elimination of surface tension will lessen or eliminate surface stiction due to liquid bridging.

In this paper, results demonstrate that supercritical carbon dioxide ($SCCO_2$) extraction can be used to remove capillary liquids (e.g. methanol) from micromachined structures, eliminating sticking caused by surface tension effects. Carbon dioxide has long been known to be a good solvent for many organic compounds (*9*), and methanol specifically is known to be very soluble in $SCCO_2$ (*10*). After the methanol has been dissolved and carried away by the supercritical fluid, the vessel is depressurized to yield dry, released microstructures. Surface tension effects have been eliminated since $SCCO_2$ has negligible surface tension like a gas. Furthermore, $SCCO_2$ possesses gas-like properties of diffusivity and viscosity (*11*) which allow the supercritical fluid to access narrow gaps under the microfeatures for removal of trapped capillary fluid. Except for approximate success or failure rates, statistical analysis of the results are not included since the intention of this paper was simply to demonstrate the feasibility of applying this technique to the manufacture of surface micromachined devices. In fact, supercritical extraction of solvent from these devices has been so successful that it has been incorporated into the micromechanics fabrication process at Sandia National Laboratories.

It should be noted that this work has been preceded by other successful applications of supercritical fluid extraction technology to eliminate surface tension effects. This includes the extraction of solvents from phase-separated polymer gels to produce microcellular foams (*12*), and the extraction of solvents from silica aerogels (*13*). Also, researchers at the University of California at Berkeley have removed solvent from micromachined samples using liquid carbon dioxide (*14*) which requires the additional step of increasing to supercritical conditions before depressurizing to avoid the sticking problems caused by a liquid/vapor interface.

Experimental

Description of pressure equipment. The extraction of methanol from micromachined samples was performed in a $SCCO_2$ pressure system, rated at 5000 psi maximum allowable working pressure, which is depicted schematically in

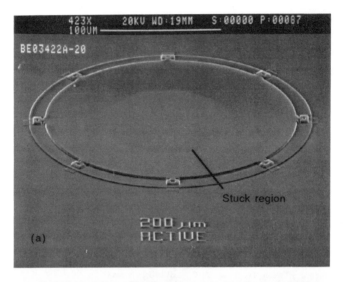

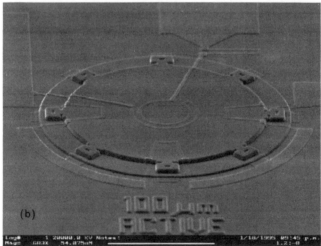

Figure 3.   A stuck 200 μm dia. pressure sensor, 423X (a), and a fully
released 100 μm dia. pressure sensor, approx. 700X (b).

Figure 4. The CO_2 source gas is supplied by three size 1A cylinders of technical grade CO_2 which is delivered through a regulator, to a pneumatic compressor (Gas Booster, model AGD-30, Haskel Corp., Burbank, CA). The compressor allows for pressurization above the critical point for CO_2. The gas flows through a 0.5 micron sintered stainless steel filter, into a one liter temperature-controlled extraction vessel (Thar Designs, Pittsburgh, PA). Temperature control is provided by a constant temperature circulating bath which flows heat transfer fluid through a water jacket surrounding the outer wall of the vessel. After leaving the extraction vessel, the CO_2 flows through a high pressure metering valve (High Pressure Equipment Co., Erie, PA) through which the pressure is reduced. The metering valve is warmed with heat tape to prevent clogging the valve with dry ice that might otherwise be formed due to Joule-Thomson cooling upon expansion of the gas. The pressure is reduced into a temperature controlled 500 ml cyclone separator recovery vessel (Thar Designs) held at 0°C. Methanol is not as soluble in gaseous CO_2 as in $SCCO_2$, so the solvent condenses and can be trapped in the separator vessel. Finally, the gas is vented out of the laboratory through a suction duct. System pressure is monitored and controlled with a modular pressure controller with digital read-out (Autoclave Engineers, Erie, PA) which controls the pneumatic compressor to maintain the desired system pressure. Over-pressure protection is provided by a 5000 psi burst disc. All tubing is 1/4 inch O.D. (0.065" wall thickness) 316 stainless steel tubing. Tube fittings are 316SS Swagelok.

The extraction vessel is made of 17-4 PH stainless steel, has an inside diameter of 3 inches and inside length of approximately 8 inches. The inside of the vessel is accessed through end caps. The vessel is supported horizontally. To minimize excess vessel volume and to support the silicon wafer pieces being extracted, two aluminum half-cylinders were machined to effectively fill the entire vessel volume. At the intersection of the half-cylinders, a $2 \, ^1/_2$ inch wide x $^3/_8$ inch high channel was machined to allow for CO_2 flow over the samples. On the bottom half-cylinder on which the wafer pieces rest, a 0.2 inch deep trough was milled to serve as a reservoir for the methanol. At the bottom of the trough are milled three rows of nine, square sample compartments that are approximately 0.775 inches on a side by approximately 0.04 inches deep which serve to keep the wafer pieces from inadvertently sliding on top of each other during the extraction. With the aluminum vessel inserts installed, the 1 liter vessel is reduced to an effective volume of about 190 ml. This arrangement forces CO_2 to flow in more direct contact with the methanol being extracted from the micromachined samples.

Experimental Procedure for Extraction of Methanol from Micromachined Structures. Surface micromachined structures were release etched, rinsed in water, then exchanged with methanol as previously described. The microstructures used in our tests were an assortment of microengines, microlevers, accelerometers, pressure sensors, and also cantilever and bridge test structures.

Control samples were allowed to air dry; that is, methanol removal was achieved through evaporation at ambient temperature. This is the method commonly used to dry micromachined parts. Test samples were supercritically extracted with carbon dioxide as follows. The micromachined samples were transfered from a methanol bath to the sample trough in the aluminum vessel insert which contained approximately 20 ml of methanol. The inserts with samples were loaded into the vessel, which was closed and then pressurized with CO_2. Pressurization typically took between 15 and 20 minutes. Extraction time was typically 30 minutes, which was empirically determined to be a sufficient amount of time to totally remove the methanol. The mass flow rate for CO_2, measured gravimetrically from the supply bottles, was approximately 30 g/min. After the

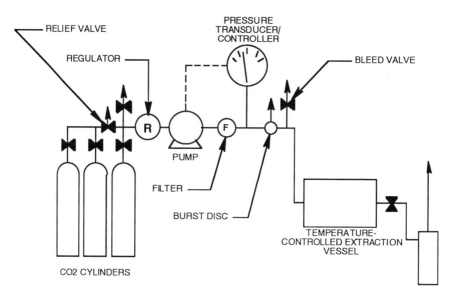

Figure 4. Schematic diagram of the 5000 psi CO_2 extraction system.

extraction, the vessel was depressurized, usually within about 15-20 minutes. The samples were then removed from the extraction vessel.

Samples were extracted under three different sets of supercritical conditions. These are 1500 psi, 40°C (ρ_{CO2}= 0.65 g/ml), 2300 psi, 40°C (ρ_{CO2}= 0.79 g/ml), and 3000 psi, 40°C (ρ_{CO2}= 0.85 g/ml). These conditions were chosen to achieve supercritical fluid densities lower than, equal to, and greater than that of methanol (0.79 g/ml). After methanol removal, control and test samples were observed and compared using scanning electron microscopy (SEM). When applicable (e.g., microengines), an electrical bias was applied in an attempt to operate the microstructures.

Results and Discussion

Comparison of Supercritical Carbon Dioxide Extraction with Conventional Air Drying. Supercritical carbon dioxide was demonstrated to reliably extract methanol from all the tested microstructures without sticking due to surface tension effects. In contrast, approximately 90% of the air dried parts were stuck overall. For some samples that were supercritically extracted, fabrication problems, such as incomplete etches, were uncovered that had previously been masked by the sticking problems experienced with conventional air drying.

Almost 100% of the SCCO$_2$ extracted microengines were free from sticking compared to only a 10% success rate for air-dried microengines. Surface stiction was observed in the springs that support the comb drive actuators. Upon sticking, the springs inhibit the movement of the comb drives, rendering the microengine useless. A micrograph comparison of a stuck air-dried microengine spring with that of a successfully released SCCO$_2$ extracted microengine is shown in Figure 5.

Similar results were seen with microlever actuators with nearly 100% success with supercritical drying as opposed to nearly 100% failure rate with air drying. The higher than average failure rate for air-dried microlever actuators is due to the exceptionally delicate structures that are even more susceptible to stiction than other, somewhat more robust devices. Figure 6 shows a micrograph of a microlever actuator with higher magnifications of a stuck air-dried lever compared to a released lever that was SCCO$_2$ extracted.

Accelerometer stiction was not visually obvious using SEM. To test for stiction, electrostatic deflection of the plate from its equilibrium position was measured. By recording the voltage necessary to deflect the plate, a relative measure of the effectiveness of the release was made. For air drying, nearly all accelerometers were stuck and did not deflect electrostatically. All accelerometers that were SCCO$_2$ extracted were free and unstuck. In subsequent deflection attempts, some of the accelerometers did show signs of sticking. This is believed to be a processing phenomenon not related to the SCCO$_2$ cleaning since the devices were fully released immediately after the SCCO$_2$ extraction. It is possible with sufficient humidity that condensation of water on the micromachined surfaces may have caused stiction after solvent removal. NH$_4$F etching before drying to slightly roughen the solid surfaces may help to increase the long term reliablility of the accelerometers. Micrographs of a supercritically extracted accelerometer are shown in Figure 7.

Pressure sensors above 200 μm in diameter were usually stuck to the substrate after a standard release and air dry. SCCO$_2$ extraction has produced freed structures up to 500 μm, the maximum size that was attempted.

Bridge and cantilever test structures were SCCO$_2$ extracted usually without surface stiction. For bridge structures, which are supported at both ends, beam lengths of up to 1.5 millimeters (mm) were fully released. For air dried bridges, most under 1 mm were released and bridges from 1 to 1.5 mm had mixed results

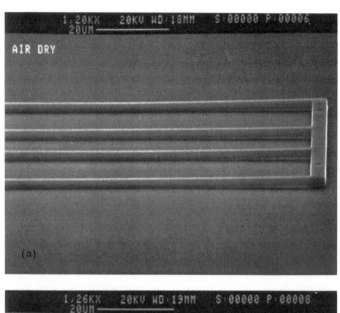

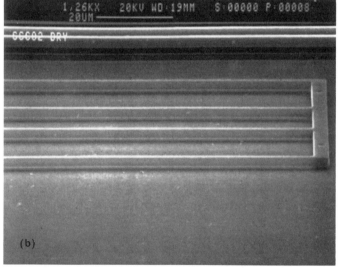

Figure 5. SEM of a stuck air-dried microengine spring, 1200X (a) and that of a SCCO$_2$ extracted spring that is fully released, 1260X (b).

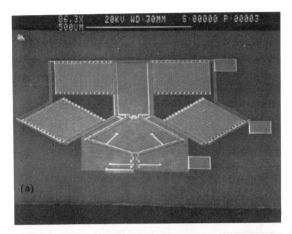

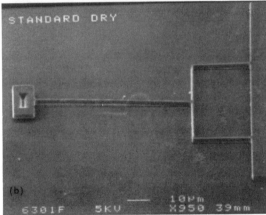

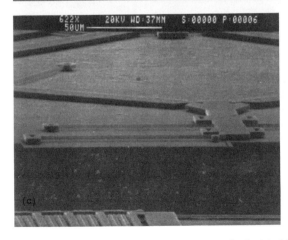

Figure 6. Micrograph of a microlever, 86.3X (a), a stuck air-dried lever arm, 950X (b), and fully released $SCCO_2$ extracted lever arms, 622X (c).

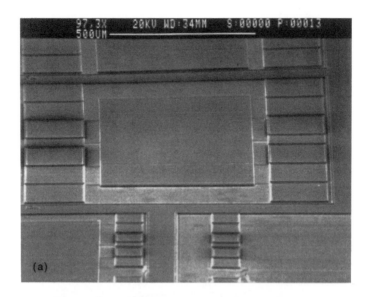

Figure 7. A micromachined accelerometer, 97.3X (a), and a close-up view of the accelerometer plate near the substrate, 2300X (b).

with some released and some stuck. Overall, $SCCO_2$ performed considerably better than air drying for bridge structures.
Because cantilever test structures are anchored only at one end, they are generally more compliant than bridges. Cantilever beam widths of 10 and 20 microns, each with 2 micron thickness were dryed using both techniques for lengths up to 500 microns. Typically, cantilevers above 150-200 microns in length were stuck for both widths using standard air drying. Virtually all cantilever beams up to 500 microns were fully released using $SCCO_2$ extraction. A comparison of air dried and $SCCO_2$ extracted cantilever test structures is shown in Figure 8. It was found that any sticking of beams observed after $SCCO_2$ drying was invariably caused by electrostatic forces generated by the electron beam in the SEM. In fact, supercritically extracted cantilever beams of 1 mm in length were observed to be consistently released when viewed with a light microscope.

Effects of Supercritical Carbon Dioxide Density on Methanol Extraction Time. Samples were successfully extracted with $SCCO_2$ at all three sets of supercritical conditions that were attempted (1500 psi, 40.0°C; 2300 psi, 40.0°C; 3000 psi, 40.0°C). There was a noticeable difference, however, in the amount of time required to remove methanol from microstructures at the lowest pressure/density conditions, 1500 psi, 40.0°C ($\rho_{CO2}= 0.65$ $^g/_{ml}$). In most cases under these conditions, the typical 30 minute extraction time was insufficient to remove all the methanol from the solvent reservoir in the vessel insert. In instances where microstructures were stuck, it was believed that surface tension effects of the remaining methanol drying from the gaps under the microstructures caused the surface stiction.
It is believed that both methanol solubility and the density difference between methanol and $SCCO_2$ affect the rate at which the solvent is carried away from the solvent reservoir. While it has already been established that methanol is very soluble in $SCCO_2$, at supercritical fluid densities lower than that of methanol, it may take longer for the $SCCO_2$ to solubilize the solvent since the lower density extraction fluid does not easily displace the standing pool of methanol. At supercritical conditions which match or exceed the methanol density, the methanol/$SCCO_2$ interface may be more easily broken and the solvent displaced and solubilized. Methanol trapped in the narrow gaps under micromachined structures can then be displaced and dissolved due to the gas-like diffusivity of the supercritical fluid. Higher, more turbulent $SCCO_2$ flow rates may be required to overcome the interfacial energy of methanol and decrease extraction times at the lower density conditions.

Conclusions

We have demonstrated that supercritical carbon dioxide extraction can be used to remove methanol to release compliant surface micromachined structures on silicon wafers. $SCCO_2$ extraction released virtually all samples of a variety of microcomponents and test structures. This is a vast improvement over conventional air drying which has a failure rate, due to surface stiction, of up to 90%.
It was also shown that extraction times of shorter duration were required when supercritical fluid densities which matched or exceeded that of the liquid solvent (e.g., methanol) were used. It is believed this occurs since a higher density extraction fluid is more effective at displacing the liquid solvent, overcoming its interfacial energy, thereby enhancing the dissolution and solvent removal properties of the supercritical fluid.
At Sandia National Laboratories, $SCCO_2$ is being used to remove solvent from surface micromachined devices on a small scale production basis. A new

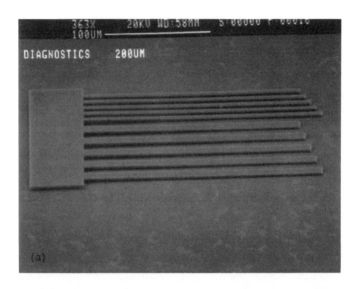

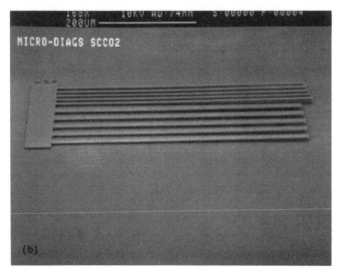

Figure 8. Air-dried cantlever test structures which have all stuck (a). The shortest air-dried cantilever is 200 μm long. SCCO$_2$ extracted cantilevers which are all released (b). The shortest SCCO$_2$ extracted cantilever is 500 μm in length.

pressure system is being designed which, when installed in the microelectronics fabrication laboratory, will be used for larger scale supercritical extraction of solvent from compliant surface micromachined silicon wafer samples.

Acknowledgments

The authors wish to acknowledge the contributions of Pat Shea who performed the SEM analysis. Micromechanical parts were supplied by William P. Eaton, Carole C. Barron, Jeff J. Sniegowski, and James H. Smith. This work was performed at Sandia National Laboratories under U.S. Department of Energy contract No. DE-AC04-94AL85000.

Literature Cited

(1)   Muller, R.S. *Sensors and Actuators*, **1990**, *A21*, 1-8.

(2)   Howe, R. T. *J. Vac. Sci. Technology*, **1988**, *6*, 1809-1813.

(3)   Peterson, K. E. *Proc. IEEE*, **1982**, *70*, 420-457.

(4)   Houston, M. R.; Maboudiank, R.; Howe, R. T. *Digest of Technical Papers, 1995 International Conf. on Solid-State Sensors and Actuators, Transducers '95*, **1995**, paper 45.

(5)   Bhushan, B. *Tribology and Mechanics of Magnetic Storage Devices*; Springer-Verlag: New York, NY, 1990; pp 231-265.

(6)   Alley, R. L.; Cuan, G. J.; Howe, R. T.; Komvopoulos, K. *Technical Digest: IEEE Solid State Sensor and Actuator Workshop*, **1992**, *Hilton Head*, 202-207.

(7)   Czarnecki, J.; Dabros, T. *J. Coll. Interf. Sci.*, **1980**, *78*, 25-30.

(8)   Yee, Y.; Chun, K.; Lee, J. D. *Digest of Technical Papers, 1995 International Conference on Solid State Actuators, Transducers '95*, **1995**, paper 44.

(9)   Francis, A. W. *J.Phys. Chem.*, **1954**, *58*, 1099-1114.

(10)   Page, S. H.; Goates, S. R.; Lee, M. L. *J. Supercritical Fluids*, **1991**, *4*, 109-117.

(11)   McHugh, M. A.; Krukonis, V. J. *Supercritical Fluid Extraction: Practice and Principles, 2nd Edition*, Butterworth-Heinemann: Boston, MA, 1994; pp 14-16.

(12)   Russick, E. M.; Aubert, J. H. *Sandia Report, Sandia National Laboratories*, **1992**, *SAND92-1122*.

(13)   Loy, D. A.; Jamison, G. M.; Baugher, B. M.; Russick, E. M.; Assink, R. A.; Prabakar, S.; Shea, K. J. *J. Non-Crystalline Solids*, **1995**, *186*, 44-53.

(14)   Mulhern, G. T.; Soane, D. S.; Howe, R. T. *The 7th International Conf. on Solid State Sensors and Actuators, Transducers '93*, **1993**, 296-299.

Chapter 19

Supercritical Fluid Extraction of Grinding and Metal Cutting Waste Contaminated with Oils

N. Dahmen, J. Schön, H. Schmieder, and K. Ebert

Institut für Technische Chemie, ITC-CPV, Forschungszentrum Karlsruhe
Gmbh, P.O.B. 3640, D–76021 Karlsruhe, Germany

The wellknown technique of supercritical fluid extraction with compressed carbon dioxide is used for an environmental application. The development of a process is described, in which residues with high oil contents from metal machining operations such as grinding or cutting can be separated in two fractions. One consists of the oil which can be recovered from the grinds in such a quality that direct reuse as grinding oil is possible; the metal fraction can be recycled by common processes. The results obtained in laboratory scale equipment were scaled up and successfully demonstrated in a pilot plant.

In Germany oil-containing residues from metal and glass working are classified as hazardous wastes. Disposal of hazardous waste requires special and costly monitoring procedures. These costs, depending on the disposal path, vary between U.S.$ 400 and 1200 per ton of grinds. According to an estimate by the German Federal Environmental Office, 150,000 to 200,000 tons of these residues are produced in Germany each year, up to 50% coming from machining with pure oils; often with oil contents up to 50%. The majority of this material arise in machining of medium and high alloyed steels or non-ferrous materials using pure grinding and cutting oils as cooling lubricants immiscible with water. In contrast to metal chips and metal powders this material consists of fine metal chips, abrasives, metal working fluids and various impurities.

At present, the following processes are available for treatment of these wastes:

- centrifugation and briquetting as pretreatment to lower the oil content to 10-20 %

- for materials with an oil content of more than 4%, thermal treatment as aggregate (additional iron) in the concrete production

- disposal by combustion with the oil acting as fuel

- for materials with an oil content of less than 4%, storing at a hazardous waste site

270

A direct recycling in metallurgical or smelting plants is not possible because the high organic contents cause off-gas problems. Here oil contents of less than 1 % are required. In Germany, several techniques are under development and testing. For example, oil separation through vacuum distillation or extraction using different organic solvents or water emulsions. The quality of the end products should permit direct recycling, since with respect to the diverse laws of waste management recycling has to be preferred to disposal. The recovered oils has to be reusable as grinding and cutting oils and the metal matrix in the respective steel and alloys producing processes.

In this paper, an extraction process in which supercritical carbon dioxide is used as solvent is described. The process was originally developed for treating glass grinds with high oil and lead contents arising in optical glass machining (*1*), but it has proved to be suitable also for treating metal grinds (*2*). The results obtained in laboratory scale equipment were scaled up and demonstrated in a pilot plant. Metal grinding sludge, also as compacts were treated successfully with supercritical carbon dioxide as extractant. The recovered oils were suitable for reuse in the respective metal machining processes.

Solubility measurements of technical oils

The oils used in metal grinding and cutting can be divided into mineral and synthetic hydrocarbon oils of different composition of their hydrocarbon components. In addition they contain a variety of additives, e.g. aging inhibitors, anticorrosives, detergents, pourpoint depressants, foam inhibitors and others. Increasingly, lubricants are used which are based on native oils like rape oil. The phase behavior of these complex mixtures was measured applying the synthetic method at temperatures from 40 to 80°C and at pressures from 10 MPa up to 40 MPa.

The experiments were carried out in a commercial phase equilibrium apparatus (Sitec PH 251-500K). The cell has a variable volume of 12 to 25 cm^3, controlled by a nitrogen-driven piston; the position of the piston is monitored by inductive measurement. Temperature in the cell was maintained constant using an electrical heating jacket. To start an experiment, a known amount of oil was filled into the cell. By adding CO_2 a homogeneous mixture was obtained and to speed up equilibration a magnetic stirrer was used. The amount of carbon dioxide was calculated from the cell volume and the density for given temperature and pressure and was determined analytically after the experiment. Then the oil was adsorbed on a charcoal filter (Supelco EnviCarb) and was determined gravimetrically, CO_2 was collected and weighed and in a gas-flask. While sampling, the pressure was kept constant by means of the piston.

To determine the phase boundary, the pressure was lowered until phase separation occurred, which was observed visually by a camera system. The pressure rates were kept constant by using a computer controlled back pressure regulator (Tescom ER2000).

In contrast to pure substances, phase separation of mixtures is a rather slow process, because the components of different molecular mass or chemical nature separate successively from the mixture. To obtain an objective measure for the phase separation, the clouding of the mixture was monitored by a device containing a photocell recording the transmission of the light from the optical fiber of the camera system. From the resulting clouding curve a starting point was chosen to characterize the beginning of the separation.

In Figure 1 the solubility of different metal working oils is presented as a function of pressure at 50°C. Depending on the composition of the oils, different solubility behaviors are observed. The lowest solubility is found for the ester-based native oil A. Large solubility differences were obtained for the two synthetic oils C and D. C is an oil of synthetic hydrocarbons and esters, containing some additives, and D is a synthetic mixture of hydrocarbons comparable to pharmaceutical white oils. For comparison, the solubility of squalane (3) (as oil representative hydrocarbon) is presented in Figure 1, showing a similar behavior as the oils B and C: an increase of pressure from 15 to 35 MPa leads to an increase of the solubility by a factor 2 to 4.

Extraction Experiments

Analysis. For analysis of the residues sum parameters were considered to be sufficient. In most cases the samples were characterized by total carbon (TC), a widely used method in environmental analysis. For TC determination 10-50 mg of material were burned within 3-5 min in an incinerator (Ströhlein C-mat 550) in an oxygen stream at 700-900°C; the CO_2 content in the off-gas stream is continuously measured with an IR detector. The time integral of the output signal is directly proportional to the carbon content of the sample. The reproducibility of the TC-values was not better than 0.25 wt%, because of the heterogeneity of the sample material. Finally the TC had to be corrected by the carbon content originating within the various steels.

Some of the samples contained considerable amounts of organics in addition to the oil, i.e. filtering aids. For these samples another method of analysis was required to prove that the carbon content remaining in a sample after extraction was not oil. Here an EPA method no. 3560 was applied using a Dionex SFE/SFC Series 600 device modified for this task. This method describes the determination of total recoverable petroleum hydrocarbons by SFE. The extract was trapped in trichlorotrifluoroethane ($C_2Cl_3F_3$); by IR-spectroscopy then the concentration of oil was determined. To calibrate the quantitative analysis, squalane was used as a representative standard oil.

Laboratory scale experiments.   The extraction parameters pressure, temperature, specific CO_2-throughput (g CO_2 / g sample material) and flow rate were studied. The laboratory scale experiments were performed in two analytical SFE devices: in a Suprex MPS 225, and an equipment using a piston metering pump with automatic pressure control (Gilson M 308 and M 821). Samples of 2-4 g of grind were treated in a 3 ml extraction cell with supercritical CO_2 flowing continuously through the cell (dynamic extraction). Frequently the extraction was interrupted and the weight reduction of the sample was determined. When constant weight was achieved, the TC value of the remaining residue was determined. After extraction the loaded CO_2 stream was depressurized through a restrictor into a solvent reservoir which trapped the precipitated oil. Depending on the solubility behavior of the extracts, precipitation or even plugging in the restrictor sometimes occurred during depressurizing. In this experiments a variable restrictor (type Suprex Duraflow) in which the cross-section of the flow could be adapted to the requirements, proved to be less troublesome than the integral fused silica restrictors, which had been used initially.

Since the oil solubility was determined using a static method and the extraction experiments were performed in a dynamic mode, the influence of the flow rate on the extraction efficiency was examined. Extraction experiments were conducted at 20 MPa and 50°C at flow rates of 0.5 to 4 g $cm^{-2}min^{-1}$. Within the experimental error, no significant influence of the flow rate on the extraction yield could be detected.

The pressure dependence of the extraction behavior was determined in the same way as described above. The weight reduction in wt% residual oil vs. specific mass flow rate is shown in Figure 2 for pressures of 15 to 30 MPa at a constant flow rate of 0.5 g cm^{-2} min^{-1} and 50°C (material no.50, Table I). In this pressure range the solvent density changes from 0.701 to 0.872 g cm^{-3}. At 30 MPa, after a specific throughput of about 13 g CO_2 /g material, more than 99.9 wt% of the oil was extracted. With the same amount of carbon dioxide at 20 MPa a value of about 9 wt% and at 15 MPa even about 27 wt% of the oil remained in the sample. The TC content achieved at constant weight was about 0.9 to 1.5 wt%. A similar extraction behavior was found for all materials tested. The specific CO_2 mass flow rate required for complete deoiling (<1 wt%) depends on the initial oil content and the nature of the matrix. Examples are presented in Table I, where the initial oil content, the specific throughput and the residual oil content achieved by the extraction are given for various materials.

The influence of temperature on the extraction yield is shown in Figure 3 at constant pressure of 20 MPa. A lower temperature of 50°C (solvent density 0.785 g cm^{-3}) leads to more efficient extraction due to the higher density of the mobile phase which, in turn, accounts for the higher solvating power of the supercritical CO_2. At the higher temperature of 80°C (0.595 g cm^{-3}) complete extraction is achieved at a considerably higher CO_2 throughput of 24 g CO_2 /g material.

Additional experiments were conducted at constant density (Figure 4). At a density of about 0.787 g cm^{-3} the higher temperature results in a better extraction behavior caused by enhanced volatility of the oil. The optimum conditions for the extraction with respect to temperature and pressure is the combination of low temperature and high pressure, as shown in Figure 4 for 50°C and 30 MPa.

Bench Scale Experiments.   A bench scale plant was used to determine whether the optimal extraction parameters obtained from the laboratory measurements (pressure, temperature, mass flow rate) were also optimal for larger scale extractions. Additionally metal compacts, which are too heterogeneous for laboratory scale experiments, were treated with supercritical carbon dioxide.

In Figure 5 a schematic flow diagram of the plant (Sitec) is shown. The storage tank contains liquid carbon dioxide, which can be compressed up to 50 MPa by a membrane pump (LEWA), producing flows up to 30 kg/h. The fluid CO_2 enters the extraction vessel of 4 dm^3 capacity either from the top or from the bottom. The process can be operated at temperatures up to 100°C. The effluent from the extractor is depressurized by a controlled needle valve. In the separator a liquid and a gas-phase are created at a pressure of about 6 MPa. The evaporating gas-phase is continuously recycled after recondensing. A concentration less than 1 ppm of hydrocarbons was found in this gas-phase, if mineral oil contaminated feed was treated. In the case of higher concentrations, the CO_2 passes an adsorption bed with activated charcoal before entering the condenser.

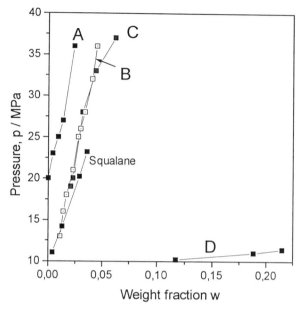

Figure 1. Solubility of metal working oils in supercritical CO_2 at 50°C A: native oil with esters, additives; B: mineral oil, with additives; C: synthetic mineral oil with synthetic esters and additives; D: synthetic mineral oil (white oil)

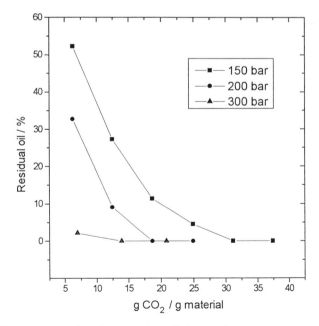

Figure 2. Variations in extraction efficiency with pressure at 50°C.

Table I. Selected materials used in the laboratory, bench and pilot plant scale experiments

Sample no.	Material	Oil-Content/%	gCO_2/gmaterial	TC/wt%
1	Metal sludge	26	15.0	< 1
2	Metal sludge	48-55	26	1.0
18	Metal sludge	20.2	14	1.3
28	Metal sludge	42.9	18	1.1
49	Metal sludge	23	27	< 35 mg/kg[a]
50	Metal sludge	50	13	< 1
53	Metal sludge	12	9-12	< 2 mg/kg[a]
I	Metal-Compact	6-7	21.1	0.43
II	Metal-Compact	—	13.1	≈ 4
III	Metal-Compact	≈3	11.4	0.15
22	Brass-Compact	2-3	10.0	0.4
23	Alu-Compact	15-16	26.8	0.1
34	Cu-Roll	3-4	11.7	< 0.2

[a] determined by EPA-method 3560

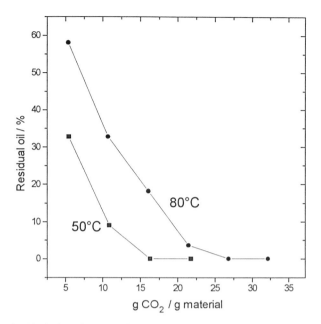

Figure 3. Variations in extraction efficiencies with temperature at 20 MPa

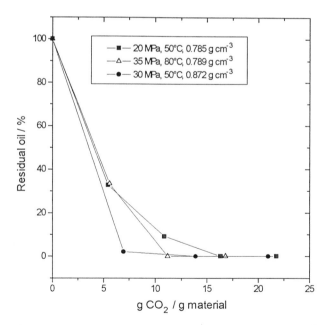

Figure 4. Extraction yield at constant density of 0.787 g cm^{-3}

In order to check the optimal extraction parameters determined in the laboratory for a scale up, certain experiments with metal grinding residues were performed in the bench scale plant. Results are presented in Figure 6 using sample no. 28 according to Table I, where the residual oil content is plotted vs. the CO_2 throughput comparing a laboratory and a bench scale experiment. Pressure and temperature were maintained constant at 30 MPa and 50°C, respectively; the flow rate in both devices was adjusted to about 2 g cm^{-2} min^{-1}. No significant differences in the extraction behavior were observed, therefore a scale up by a factor of about 1000 was justified with respect to the extraction parameters. Possibly this is not valid for other substances under test, due to the different composition and consistence of the various materials.

In the bench scale device also metal compacts were extracted. The mechanically compressed blocks consist of metal-chips and max. 40 wt% to guarantee mechanical stability. The metal compacts under test varied in composition; in addition to ferrous materials (Table I, sample no. I-III) also nonferrous samples were studied. Representative samples for the TC-analysis were taken from drill cores. The compacts were treated using a static/dynamic method, in which after stopping the CO_2-stream for a certain time span, the oil was flushed with supercritical carbon dioxide. Results are presented in Table I. For deoiling of the compacts at least 10 to 27 g CO_2 / g material were necessary. The remaining TC content was less than 1 %, with the exception of sample II, where undefined organic material was found in addition to the oil, resulting in a higher TC-value of about 4 %.

Pilot plant experiments. To demonstrate the scale-up of laboratory and bench scale parameters up to industrial dimensions, experiments in an industrial plant of Messer-Griesheim in Krefeld, Germany, were performed. The plant consists of two 200 l extraction vessels continuously flushed with supercritical carbon dioxide. In two separation vessels of 20 and 50 dm^3 volume, the loaded effluent CO_2 flow was depressurized totally to gas and the oil was precipitated. Since the flow rate in the laboratory scale experiments was found to have no effect on the extraction efficiency, the runs were performed with high flow rates of about 800 to 1200 kg/h at pressures up to 40 MPa. Because of the large amount of oil, the extraction had to be interrupted from time to time to discharge the oil from the recovery vessels.

Several batches of three different materials, no. 49, 50 and 53 according to Table 1, were tested at pressures between 30 and 40 MPa. In all cases the optimal extraction parameters, obtained from the laboratory experiments, were found to be valid also in a plant of industrial size. At the end of the extraction samples for analysis were collected from the top, the middle and the bottom of the residue. Results are given in Table I. Because sample no. 49 and 53 contained additional organic material, the completeness of the oil extraction was demonstrated by the EPA-method described above. Only inconsiderable residual oil contents in the range of mg/kg were found.

Conclusion

In the experiments described above, the extraction process was monitored discontinuously by weighing the samples or the amount of extracted after each extraction step. This is not reasonable when thinking of technical applications of this process. In regard of the uncertainty of the specific carbon dioxide throughput,

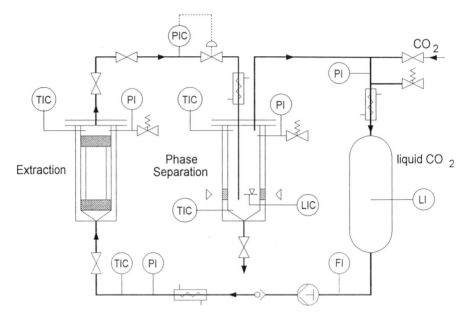

Figure 5. Flow diagram of the bench scale plant

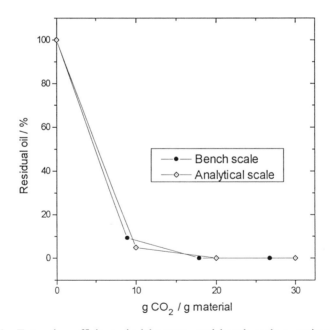

Figure 6. Extraction efficiency in laboratory and bench scale experiments at 30 MPa and 50°C

depending on the heterogeneity of the materials, and scale up effects, monitoring of the extraction is necessary to optimize extraction time. Thus an in-line monitor, based on a FID (flame ionization detector), was developed and successfully tested in a laboratory device (*4*).

The pilot plant experiments were not only performed to obtain information about scale-up effects, but also to make possible a reliable cost estimate. Based on the pilot plant experiments (plant capacity 500 t/a) the percentage of the operating costs such as for energy (current: 0.56 kWh/kg, heating: 0.35 kWh/kg) and CO_2 consumption (0.2 kg/kg) and maintenance were estimated to about 20%, staff costs were calculated to be 60-80% of the overall costs. These are very sensitive on the mass throughput. For a hypothetical plant of 2000 t/a capacity and the same conditions of extraction, a cost reduction by nearly 50% is expected. Consequently, with regard to the oil bonus arising from reusing the extracted oil and by avoiding the costs of disposal, this process is of economical interest. The absolute costs of course depend on the individual staff costs, different costs of installations and rate of utilization. The experiments described above demonstrate that extraction operating parameters determined in laboratory devices can be directly transferred to an industrial scale of operation.

Acknowledgments

The authors would like to thank Mrs. P. Kunz, Mr. Buchmüller, Mr. Schwab and Mr. Wilde for the many carefully conducted extraction experiments and analysis, the industrial partners for delivering samples of material, and Messer Griesheim for their collaboration in performing the pilot plant runs.

References

1.   Schön, J.; Dahmen, N.; Schmieder, H.; Ebert, K. *Sep. Sci. Tech.*; in press

2.   Schmieder, H.; Schön, J.; Wiegand, G.; Dahmen, N.   In *Chemistry under Extreme or Non-classic Conditions*; van Eldik, R.; Hubbard, C.D., Ed.; Spektrum Akademischer Verlag: Heidelberg, Germany, in press

3.   Schmitt, W.J.; Reid, R.C. *Chem. Eng. Comm.* **1988**, *65*,155

4.   Schön, J.; Dahmen, N. Abstracts of the 3rd European Symposium on Analytical SFC and SFE and 6th Int. Symp. on SFC and SFE, September 6.-8. 1995, Uppsala, Sweden, P 16

Chapter 20

Supercritical Fluid Extraction for Remediation of Contaminated Soil

M. Reza Ekhtera[1], G. Ali Mansoori[1,4], Michael C. Mensinger[2], Amir Rehmat[2], and Brunie Deville[3]

[1]Department of Chemical Engineering, University of Illinois at Chicago, 810 South Clinton Street, Chicago, IL 60607–7000
[2]Institute of Gas Technology, 1700 South Mount Prospect Road, Des Plaines, IL 60018
[3]U.S. Environmental Protection Agency, 26 West Martin Luther King Drive, Cincinnati, OH 45268

The supercritical fluid extraction liquid phase oxidation (SELPhOx) process is being developed as a highly flexible means of remediating and destroying both high and low concentrations of light aliphatic to heavy aromatic contaminants from solid matrices. The process employs two distinct technologies: extraction of organic contaminants with supercritical carbon dioxide and wet air oxidation (WAO) destruction of the extracted contaminants. A separation step links the two process stages. Supercritical fluid extraction tests are conducted over wide ranges of temperature, pressure, and CO_2/contaminant ratios with soils from a wood treatment plant and two manufacturing gas plant sites. Extraction of polynuclear aromatic hydrocarbons (PAHs) from these soil samples are studied experimentally. The addition of methanol as an extraction modifier was also explored. At comparable CO_2-to-contaminant ratios and extraction conditions of 48 °C and 137 atm, the total PAHs removed from the three soils ranged from 76.9 to 97.9 percent with CO_2 alone and from 88.4 to 98.6 percent with methanol added. Results of these tests are presented and analysed. A skid-mounted Field Test Unit (FTU) based on the laboratory bench-scale test results is being constructed which allows on-site testing of the integrated SELPhOx process with contaminated soils.

Supercritical fluid extraction (SFE) relies on the unique physical properties of supercritical fluids (SCF). A supercritical fluid is a compound at conditions exceeding

[4]Corresponding author

its critical temperature and pressure. SCFs have viscosities and diffusivities between liquids and gases and densities close to those of liquids. They have the solution characteristics of liquids with better mass transfer capabilities. Changes in the pressure (and as a result density) of a SCF can be used to change its solvation power[1]. For example, the critical conditions for carbon dioxide are 31 °C and 75 atm. At 38 °C, naphthalene solubility in CO_2 increases 20-fold from 0.1 to 2.0 weight percent by increasing the pressure from 72.4 atm (below the critical point) to 92.9 atm (above the critical point)[2].

Supercritical carbon dioxide is used for extraction in the SFE stage of the SELPhOx process because of its low cost, environmental acceptability, and high solvation power. Many organic compounds are soluble in supercritical CO_2, but solubilities vary greatly for different compounds and under different conditions. Alkanes, up to C_{12}, are completely miscible in supercritical CO_2. As a general rule, solubility decreases with increasing molecular size and with the addition of branching, aromatic groups, hydroxyl groups, and halogen atoms to the compound[3]. Extraction of partially soluble compounds can often be enhanced by increasing the temperature or pressure. Research has shown that organic pollutants as diverse as phenol, pesticides, PCBs, dioxines, and coal tar residues can be extracted from soils using supercritical CO_2[4,5,6]. Extraction of recalcitrant compounds can be improved with the addition of an extraction modifier, such as an alcohol, to the SCF[7].

Extracted contaminants are destroyed in the SELPhOx process by wet air oxidation. WAO is a thermal treatment process in which a slurry of water and carbonaceous material is heated to 175 to 345 °C (below the 374.2 °C critical temperature of water) under pressure and in the presence of oxygen. Oxygen can be supplied in air, enriched air, pure O_2, or an oxidizing chemical such as hydrogen peroxide. With sufficient oxygen, the carbonaceous material can be oxidized completely to CO_2 and water. Typically, the oxidation provides a portion of the process thermal energy requirement.

Wet air oxidation has been commercially practiced for several decades with plants around the world treating municipal and industrial wastes. Zimpro Environmental, Inc. (ZEI) uses a vertically-oriented, cylindrical reactor through which air or oxygen flows upward co-currently with the fluid being treated. ZEI has provided over 200 full-scale WAO units and is the major supplier of this technology. The WAO process has been applied successfully to a variety of wastes including municipal sludge, acrylonitrile, explosives, and a wide range of petrochemicals[8]. Bench-scale WAO tests conducted at temperatures of 250 to 320 °C for 60 to 120 minutes (with or without catalyst) showed greater than 99 percent destruction of many compounds. Halogenated aromatic compounds without other non-halogen functional groups (chlorobenzene, PCBs, etc.) proved to be more recalcitrant with conventional WAO destructions of 63 to 85 percent[9,10,11]. Catalysts and stronger oxidizing agents can be used to achieve more complete destruction of these compounds.

The proposed supercritical fluid extraction joined with liquid phase oxidation (SELPhOx) process seems to have a number of advantages over other processes for remediating Superfund site soils contaminated with organic compounds[12]. SFE with CO_2 does not destroy the humic content of the soil. A clean soil can be returned to the site. Unlike incineration, SELPhOx generates no hazardous compounds such as dioxins and furans. Gaseous emissions from the WAO stage are primarily CO_2 which can be recycled to the SFE stage. Contaminant removal and destruction levels can exceed 99 percent.

Experimental

Laboratory-Scale Batch Tests. In the experimental laboratory scale batch tests performed the extraction efficiency of polynuclear aromatic hydrocarbons (PAHs) from various soil samples is studied. In Figure 1 the structure, number, name, formula and molecular weights of 16 poly aromatic hydrocarbons studied in the course of this research project are reported.

The supercritical fluid extraction stage of the SELPhOx process was evaluated through a series of experiments in a batch supercritical fluid extraction (SFE) unit (LDC Analytical, Inc.) modified. A schematic diagram of the unit is shown in Figure 2. Three type 316 stainless steel vessels, a 55-ml extraction cell and two 150-ml separation vessels, are connected in series and have individual pressure regulators and heaters. A metering pump with head chiller supplies extraction fluid to the system.

System additions include fine screens at the top and bottom of the extraction cell, an exit gas dry test meter, and an exit line activated carbon filter to capture any contaminants not collected in the separation solvent. Dip tubes added to the separation vessels improve contaminant collection. Temperature measurement accuracy was improved by moving thermocouples from the top to the middle of each vessel.

Before each test, a weighed 30-gram sample of contaminated soil was placed in the extraction cell, and 50 ml of methylene chloride was charged to each separation vessel. Methylene chloride was used to ensure high contaminant collection in a medium which would allow concentration in methanol at low temperature with no loss through volatilization. Once charged, the vessels were raised to the desired pressures. The extraction cell was then heated to the test condition. A steady SCF flow was then established through the three vessels, and the test was continued until the desired fluid to contaminant ratio was achieved. The unit was then cooled and depressurized. Residual soil contaminant concentrations were determined by acetone:hexane soxhlet extraction followed by GC analysis. The entire unit was washed with methylene chloride after each test.

The first soil tested, Soil 1, was collected by Biotrol, Inc. from the North Cavalcade Superfund site in Texas, a former wood treating plant[13]. The other two soils were collected from manufactured gas plant (MGP) sites. Soils were riffled and refrigerated in air- and light-tight containers until needed for testing. All three soils contain high concentrations of 2- to 6-ring polynuclear aromatic hydrocarbon (PAH) compounds.

Field Testing

A skid-mounted Field Test Unit (FTU) was constructed which allows on-site testing of the integrated SELPhOx process with contaminated soils. A schematic diagram of the FTU is shown in Figure 3. The SFE stage is operated in semi-continuous mode with soil batches of 1 to 5 kg. The wet air oxidation (WAO) stage

	1	Naphthalene	$C_{10}H_8$	128.74
	2	Acenaphtylene	$C_{12}H_8$	152.21
	3	Acenaphtene	$C_{12}H_{10}$	154.21
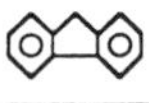	4	Fluorene	$C_{13}H_{10}$	166.22
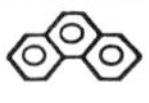	5	Phenanthrene	$C_{14}H_{10}$	178.23
	6	Anthracene	$C_{14}H_{10}$	178.23
	7	Fluoranthene	$C_{16}H_{10}$	202.55
	8	Pyrene	$C_{16}H_{10}$	202.55
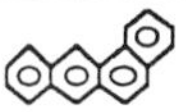	9	Benz(a)anthracene	$C_{18}H_{12}$	228.29
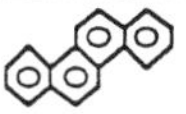	10	Chrysene	$C_{18}H_{12}$	228.29
	11	Benzo(b)fluoranthene	$C_{20}H_{12}$	252.2
	12	Benzo(k)fluoranthene	$C_{20}H_{12}$	252.2
	13	Benzo(a)pyrene	$C_{20}H_{12}$	252.2
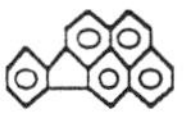	14	Indeno(1,2,3-cd)pyrene	$C_{22}H_{12}$	276.2
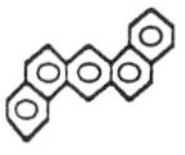	15	Dibenz(a,h)anthracene	$C_{22}H_{14}$	278.2
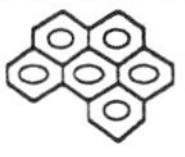	16	Benzo(ghi)perylene	$C_{22}H_{12}$	276.2

Figure 1: Structure, number, name, formula and molecular weights of poly aromatic hydrocarbons (PAHs)

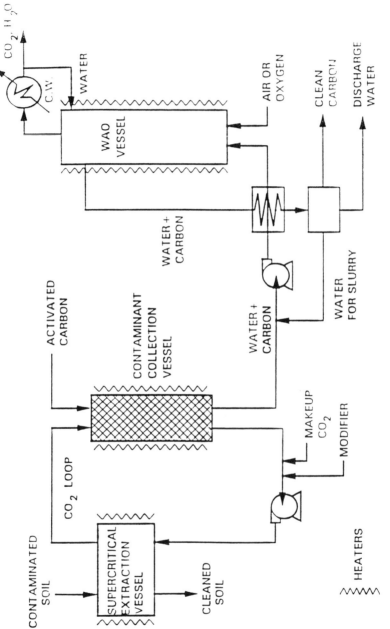

Figure 2: Schematic diagram of the supercritical fluid extraction unit

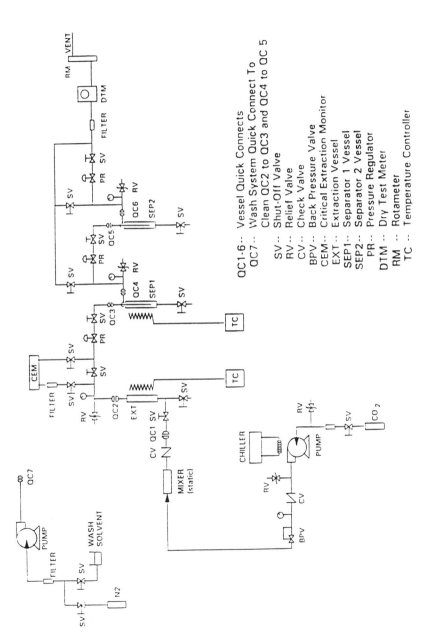

QC1-6 -- Vessel Quick Connects
QC7 -- Wash System Quick Connect To
        Clean QC2 to QC3 and QC4 to QC 5
SV -- Shut-Off Valve
RV -- Relief Valve
CV -- Check Valve
BPV -- Back Pressure Valve
CEM -- Critical Extraction Monitor
EXT -- Extraction Vessel
SEP1 -- Separator 1 Vessel
SEP2 -- Separator 2 Vessel
PR -- Pressure Regulator
DTM -- Dry Test Meter
RM -- Rotameter
TC -- Temperature Controller

Figure 3: Schematic diagram of the Field Test Unit

runs continuously with the aqueous separation solvent as feed. The flexible design incorporates CO_2 recycle in the SFE stage, CO_2 separation from the WAO stage effluent, and means for CO_2 reuse. The separation stage is an activated carbon. The data collection and display of the FTU is by a dedicated on-line computer system.

Discussion

The effectiveness of the SFE process was determined by following the fate of sixteen compounds identified by EPA method SW-846, 8100.[15] Concentrations of other aromatic, alkylaromatic, and non-aromatic hydrocarbons present were not determined. No aliphatic compounds were selected for analysis because aromatic contaminants predominate in these soils and because aliphatic compounds are easier to extract with supercritical CO_2.

Soil characterization data and PAH concentrations in the three soils are presented in Table 1. Soil 1 is characterized as a sandy loam; the other two soils are sandy. Total contaminant concentrations are similar at 1500 to 2000 µg/g, but contaminant distributions are different. PAHs with 4 to 6 rings comprise only 38 percent of the contaminants in the wood treating site soil. Soils 2 and 3 from MGP sites contain heavier compounds with 63 and 51 percent of PAHs having 4 to 6 rings. Two of the soils (1 and 3) are neutral while Soil 2 is acidic with a pH of 4.5.

Twenty-five SFE tests were conducted including eleven with Soil 1 and seven with each of the other soils. Summaries of the test conditions and extraction results are provided in Tables 2, 3, and 4. The first extraction test with each soil was at a baseline condition of 47 °C and 137 atm, using a CO_2-to-contaminant ratio of 7 kg/g. Soil 1 extraction conditions were individually varied to cover a temperature range of 33 to 77 °C, a pressure range of 76 to 273 atm, a CO_2-to-contaminant range of 1.7 to 13.3 kg/g, and the addition of 5 percent methanol as a co-solvent modifier. In tests with soils 2 and 3, pressures of 137 and 273 atm were used, and the CO_2-to-contaminant ratio was raised to as high as 67 kg/g. Temperature was not varied. The 5 percent methanol modifier was used in two tests with each of these soils.

Test No. E-3, E-7, and E-8 were conducted with Soil 1 at similar conditions with different extraction temperatures. Extraction levels increased as the temperature was raised from 33 °C, just above the critical temperature, to 48 °C. The extraction level was unchanged for naphthalene but increased with increasing molecular size. The extraction of 5- and 6-ring PAHs increased from 44 to 58 percent. A further temperature increase to 77 °C resulted in a decrease in extraction levels with the largest effects again seen for the largest compounds. Extraction of 5- and 6-ring PAHs decreased from 58 to 1 percent.

The extraction of PAH contaminants depended strongly on the extraction conditions, the molecular size of the contaminants, and the soil type. The effects of pressure, CO_2-to-contaminant ratio, and an extraction modifier on the extractions from Soil 1 and shown in Table 2.

Total PAH extraction increased with increasing pressure and with increasing CO_2-to-contaminant ratio. PAH extractions from Soil 1 were similar at pressures of 76, 103, and 137 atm, but were higher at 237 atm. Total residual PAHs were 121 µg/g at 137 atm and 77 µg/g (35 percent lower) at 237 atm. Increasing the extraction pressure from 137 to 273 atm for Soil 2 also increased total PAH removal, but the residual PAH

Table 1. SOIL CHARACTERISTICS AND CONTAMINANT CONCENTRATIONS

Compound	Number of Rings	Concentration in Soil, μg/g		
		Soil 1	Soil 2	Soil 3
Naphthalene	2	50	87	292
Acenaphthylene	3	7	26	47
Acenaphthene	3	232	44	171
Fluorene	3	199	58	107
Phenanthrene	3	603	262	248
Anthracene	3	91	70	97
Fluoranthene	4	311	210	156
Pyrene	4	244	279	288
Benzo(a)anthracene	4	62	97	88
Chrysene	4	53	104	113
Benzo(b+k)fluoranthenes	5	43	100	118
Benzo(a)pyrene	5	18	81	103
Indeno(123-cd)pyrene + Dibenzo(ah)anthracene	6.5	6	53	71
Benzo(ghi)perylene	6	7	33	70
2 Ring		50	87	292
3 Ring		1132	461	669
4 Ring		670	689	646
5+6 Ring		73	267	362
Total		1925	1504	1969
pH		7.9	4.5	7.1
Total Organic Matter, wt %		0.7	2.5	1.7
Phosphorus, μg/g		5	17	9
Potassium, μg/g		55	55	85
Nitrate Nitrogen, μg/g		4	< 1	1
Soil Type, wt %				
Sand		59	96	85
Silt		27	1	11
Clay		14	3	4

Table 2. CONTAMINANT EXTRACTION FROM SOIL 1
Initial PAH concentration - 1925 μg/g

Test	E-3	E-4	E-5	E-6	E-7	E-8
Temperature, °C	45	44	42	44	33	77
Pressure, atm	137	103	76	273	137	137
Co-solvent (5% MeOH)	No	No	No	No	No	No
CO_2/contaminant, kg/g	7.6	7.7	6.6	6.8	6.5	6.7
Contaminant Removal, %						
Naphthalene	98.7	97.8	99.2	99.4	99.2	98.0
Phenanthrene	98.0	97.9	97.1	98.1	95.6	94.2
Benzo(a)pyrene	68.7	63.3	69.4	77.2	62.1	6.2
2-Ring Compounds	98.7	97.8	99.2	99.4	99.2	98.0
3-Ring Compounds	97.3	97.4	96.7	97.4	95.7	89.4
4-Ring Compounds	90.9	91.9	89.8	96.2	86.4	69.5
5-6-Ring Compounds	58.2	53.5	52.1	70.3	43.8	1.2
Total Compounds	93.7	93.8	92.6	96.0	90.6	79.3
Residual PAH, μg/g	121	119	142	77	181	398

Test	E-9	E-10	E-11	E-12	E-13
Temperature, °C	40	43	46	44	42
Pressure, atm	137	137	137	137	273
Co-solvent (5% MeOH)	No	No	No	Yes	Yes
CO_2/contaminant, kg/g	1.7	4.5	13.3	7.1	7.0
Contaminant Removal, %					
Naphthalene	98.2	98.9	99.2	98.2	98.5
Phenanthrene	90.3	93.7	98.4	97.8	98.8
Benzo(a)pyrene	44.7	34.0	47.8	91.8	78.7
2-Ring Compounds	98.2	98.9	99.2	98.2	98.5
3-Ring Compounds	91.5	93.4	97.6	97.5	98.6
4-Ring Compounds	75.5	79.7	90.4	97.2	97.1
5-6-Ring Compounds	40.6	21.2	40.9	90.6	84.3
Total Compounds	84.2	86.0	93.0	97.2	97.5
Residual PAH, μg/g	304	269	135	54	48

Table 3. CONTAMINANT EXTRACTION FROM SOIL 2
Initial PAH concentration - 1504 µg/g

Test	E-28	E-30	E-35	E-29	E-34	E-37	E-36
Temperature. °C	49	47	47	48	48	47	48
Pressure. atm	137	137	137	273	273	137	273
Co-solvent (5% MeOH)	No	No	No	No	No	Yes	Yes
CO_2/contaminant. kg/g	16.2	31.7	67.5	14.9	49.1	63.6	46.8
Contaminant Removal. %							
Naphthalene	11.0	26.3	69.5	16.9	52.5	88.6	85.0
Phenanthrene	84.6	90.6	92.3	89.7	92.0	96.4	96.2
Benzo(a)pyrene	40.0	31.3	37.7	39.5	44.8	67.4	79.7
2-Ring Compounds	11.0	26.3	69.5	16.9	52.5	88.6	85.0
3-Ring Compounds	80.2	84.3	92.9	84.3	92.3	96.2	96.2
4-Ring Compounds	63.6	75.3	83.9	76.4	87.2	93.6	95.3
5-6-Ring Compounds	17.2	25.2	33.8	27.2	38.5	61.5	75.5
Total Compounds	57.4	66.4	76.9	66.6	78.1	88.4	91.5
Residual PAH. µg/g	641	505	347	502	329	174	128

Table 4. CONTAMINANT EXTRACTION FROM SOIL 3
Initial PAH concentration - 1969 µg/g

Test	E-31	E-33	E-41	E-32	E-40	E-38	E-39
Temperature. °C	48	48	47	48	48	47	47
Pressure. atm	137	137	137	273	273	137	273
Co-solvent (5% MeOH)	No	No	No	No	No	Yes	Yes
CO_2/contaminant. kg/g	12.0	25.5	51.1	12.4	37.7	51.2	37.9
Contaminant Removal. %							
Naphthalene	99.2	99.3	99.4	99.3	99.4	99.6	99.6
Phenanthrene	99.8	98.7	98.6	98.7	98.5	99.0	98.9
Benzo(a)pyrene	93.9	94.8	94.7	92.1	95.7	97.8	98.4
2-Ring Compounds	99.2	99.3	99.4	99.3	99.4	99.6	99.6
3-Ring Compounds	98.9	98.8	99.0	99.0	99.0	99.1	99.2
4-Ring Compounds	98.1	98.2	98.1	96.0	97.7	98.4	99.1
5-6-Ring Compounds	93.8	94.1	94.2	91.2	95.2	97.5	98.5
Total Compounds	97.8	97.8	97.9	96.6	97.9	98.6	99.1
Residual PAH. µg/g	43	43	41	67	41	28	18

concentration was reduced by approximately 20 percent. No pressure effect was observed for Soil 3 extraction. This may be due to the already high extraction levels (exceeding 98 percent of total PAHs) achieved for this soil.

Extraction levels increased with increasing CO_2-to-contaminant ratios. Soil 1 extraction levels increased with increasing CO_2 to contaminant ratios up to 7 kg/g, but higher loadings did not increase extraction. Soil 2 extraction levels were lower, and increased over the full range of CO_2-to-contaminant ratios used. Extraction levels from Soil 3 were very high (97 to 98 percent) and did not vary with changes in the solvent-to-contaminant ratio. Removal of contaminants was highest for 2- and 3-ring compounds, decreasing with increasing molecular size. The reason for the low Soil 2 naphthalene extraction is unclear. Extraction data for 5- and 6-ring PAHs is scattered, reflecting low concentrations approaching the method detection limit.

The effects of a 5 percent methanol co-solvent modifier at 137 and 273 atm are shown in Tables 2, 3 and 4 (Soils 1, 2 and 3). Residual PAH levels were reduced by approximately 50 percent for all three soils compared with extraction with CO_2 alone. At comparable CO_2-to-contaminant ratios and extraction conditions of 48 °C and 137 atm, the total PAHs removed from Soils 1, 2, and 3 were 93.7, 76.9, and 97.9 percent with CO_2 alone and 97.2, 88.4, and 98.6 percent with methanol added. Adding methanol decreased the PAH level after extraction from 121 to 53 µg/g for Soil 1, from 347 to 174 µg/g for Soil 2, and from 41 to 28 µg/g for Soil 3. The methanol co-solvent had the largest effect on extraction of the largest PAH compounds which were the most recalcitrant with CO_2 alone.

The three soils studied are contaminated with similar levels of PAH compounds. The soils, however, are from different locations, and the sources and distributions of hydrocarbons in the soils are different. Extraction levels from the soils decreased from Soil 3 to Soil 1 to Soil 2. Soils 2 and 3 are almost entirely sand while Soil 1 is sandy with 14 percent clay. The clay in Soil 1 did not appear to interfere with the extraction of PAH contaminants.

Table 5 presents results of blank and spike tests with sand and Soil 1. Extraction of a clean sand found a PAH level of 4 ppm, an artifact of the analysis. Spike 1 with naphthalene and 2 ethyl-naphthalene was completely extracted from the sand. Spike 2 with the 16 compounds under study and 2 ethyl-naphthalene was extracted at comparable levels from the sand and Soil 1. This confirms that a clay content of 14 percent does not interfere with the extraction of PAH contaminants.

Conclusions

Supercritical CO_2 has been shown to successfully extract 2- to 6-ring PAH contaminants from soil at low temperatures and moderate pressures. High clay content soils which are difficult to remediate by other methods may respond well to supercritical CO_2 extraction since the 14 percent clay content of soil 1 did not hinder extraction. Extraction was highest at 60 °C, decreasing with both lower and higher temperatures using 5% methanol in CO_2. An optimum extraction temperature above, but near, the critical temperature agrees with compiled data showing naphthalene solubility in supercritical CO_2 decreases with increasing temperature at pressures below 120 atm.

Extraction levels increase with increasing pressure and increasing CO_2-to-contaminant ratios. Removal of PAHs decreases with an increasing number of rings in the compound. This reflects the decreasing volatility of the larger compounds.

Table 5. BLANK AND SPIKE EXTRACTION TEST RESULTS

Test	E-14	E-15	E-16	E-17
Soil	Sand	Sand	Sand	Soil 1
Spike	Blank	Spike 1	Spike 2	Spike 2
Temperature, °C	43	46	47	44
Pressure, atm	137	137	137	137
CO_2/contaminant, kg/g	--	64.2	6.7	3.4
Contaminant Removal, %				
Naphthalene	--	--	99.7	99.6
2 Ethyl-naphthalene	--	99.9	99.9	98.3
Acenaphthylene	--	99.9	99.9	98.6
Phenanthrene	--	--	97.1	97.8
Benzo(a)pyrene	--	--	92.2	74.5
2-Ring Compounds	--	99.9	99.8	99.1
3-Ring Compounds	--	99.9	99.0	97.7
4-Ring Compounds	--	--	97.4	91.7
5-6-Ring Compounds	--	--	68.4	51.6
Total Compounds	--	99.9	90.5	89.4
Charged PAHs, μg/g	0	200	1578	3504
Residual PAHs, μg/g	4	4	150	371

Blank   - No contaminants added
Spike 1 - 100 μg/g of 2 ethyl-naphthalene and acenaphthylene
Spike 2 - 100 μg/g of 2 ethyl-naphthalene and 16 PAHs analyzed

Adding methanol as a co-solvent increases PAH extraction by enhancing the SCF solvation power. The greatest effect is on the removal of large compounds because they have the lowest solubility in CO_2. Extraction levels of 78 to 98 percent were obtained using CO_2 alone. With methanol added, higher extraction levels of 92 to 99 percent were obtained.

Significant differences were obtained in extraction levels from different soils. This has been observed previously with extraction decreasing with increasing soil organic carbon content and higher contaminant-to-organic matter affinity[4]. This could account for the comparatively low extraction levels from Soil 2, the soil with the highest organic matter content. The extraction levels from Soils 1 and 3 are not in the same order as the soil organic matter content. Soil 1 had the highest silt and clay contents which may adversely affect contaminant extraction. Also, Soil 1 was from a wood treating site with a different mix of PAHs than Soils 2 and 3 from MGP sites. More investigation is required to explain the differences in extraction levels from different soils.

In order to illustrate the effect of various variables on the extraction Figures 4-8 for Soil 2 are reported here. According to Figure 4 the total extraction efficiency of PAHs decreases with time sharply at the beginning of the process. When we look at the extraction efficiency of individual PAHs versus time (Figure 5) the same trend as in figure 4 is noticed. However, PAH no.s 7 and 8 have the highest efficiency at all times. In Figure 6 we report the effect of extraction time on total percent removal of PAHs from Soil 2. According to this figure the total percent removal of PAHs from Soil 2 increases exponentially at the beginning of the extraction, and it levels off asymptotically in the later parts of the process. The effect of concentration of methanol on the extraction of PAHs from Soil 2 are demonstrated on Figure 7. According to this figure the most pronounced effect on extraction is at 5% methanol for PAH no.'s 7 and larger. According to Figure 8 with 5% methanol in solvent and for the four temperatures of operation, T= 60 °C extraction has the highest level of removal of all PAHs (except PAH no. 3) than the 45, 100, and 140 °C.

Nomenclature and Abbreviations

CO_2	carbon dioxide
EPA	U.S. Environmental Protection Agency
FTU	field test unit
MGP	manufactured gas plant
P	absolute pressure
PAH	polynuclear aromatic hydrocarbon
ppm	parts per million
SCE	supercritical extraction
SCF	supercritical fluids
SELPhOx	Supercritical Extraction/Liquid-Phase Oxidation
T	absolute temperature
WAO	wet air oxidation

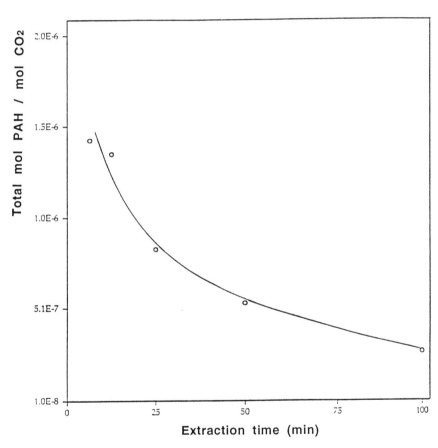

Figure 4: Effect of extraction time on the total extraction efficiency of PAHs from Soil 2 (extraction condition: 4 ml/min, 45 °C, 137 atm)

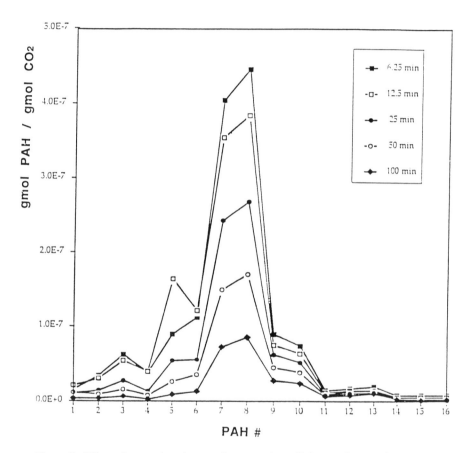

Figure 5: Effect of extraction time on the extraction efficiency of PAHs from Soil 2
(Extraction conditions: 4 ml/min, 45 °C, 137 atm)

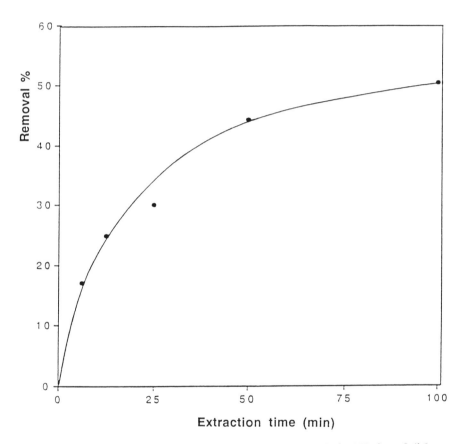

Figure 6: Effect of extraction time on total percent removal of PAHs from Soil 2
(Extraction conditions: 4 ml/min, 45 °C, 137 atm)

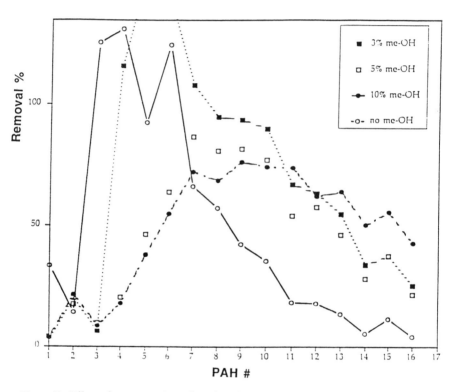

Figure 7: Effect of concentration of methanol on the extraction of PAHs from Soil 2
(Extraction conditions: 4 ml/min, 45 °C, 137 atm)

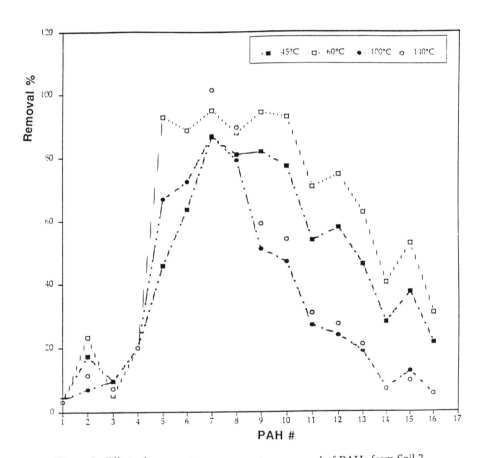

Figure 8: Effect of temperature on percentage removal of PAHs from Soil 2
(Extraction conditions: 4 ml/min, 137 atm, 5% methanol in CO_2)

Acknowledgements: The authors would like to thank Mr. H. Touba and Prof. M. Edalat for their help in testing the SELPhOx Field Test Unit. This research is supported in part by the U.S. EPA SITE program under the Environmental Protection Agency Assistance Agreement No. CR-822701-01-1 and in part by the IGT Sustaining Membership Program.

Literature Cited

1. Kwak, T.Y.; Mansoori, G.A. *Chemical Eng. Science* 1986, vol. 41, pp.1303-1309.
2. Mansoori, G.A.: Schulz; K.; Martinelli, E. *BIO/TECHNOLOGY* 1988, (Feature Article) vol. 6, pp.393-396.
3. Francis, A. *J. Phvs.-Chem.* 1954, vol. 58, pp.1099-1114.
4. Brady, B.; Kao, C.; Dooley, K.; Knopf, F. *Ind. Eng. Chem. Res.* 1987, vol. 26, pp. 261-268.
5. Bartle, K.D.; Clifford, A.A.; Jafar, S.A. *J. Phys. Chem. Ref. Data* 1991, vol. 20, pp. 713-757.
6. Wright, B.; Wright, C.; Fruchter, J. *Energy and Fuels* 1989, vol. 3, pp. 474-480.
7. Kwon, Y.J.; Mansoori, G.A. *J. Supercritical Fluids* 1993, vol. 6, pp. 173-180.
8. Schaefer, P. *Hydrocarbon Proc.* 1981, October, pp. 100-104.
9. Foussard, J.; Debellefontaine, H.; Besombes-Vailhe, J. *J. Envir. Eng.* 1989, vol. 115, pp.367-373, .
10. Joglekar, H.; Samant, S.; Joshi, J. *Water Res.* 1991, vol. 25, pp. 135-142,.
11. Mishra, V.; Mahajani, V.; Joshi, J. *Ind. Eng. Chem. Res.* 1995, vol. 34, pp. 2-48.
12. Ekhtera, M. R.; Mansoori, G.A.; Mensinger, M.C.; Rehmat, A.; Deville, B. "Design and Construction of a Supercritical Fluid Extraction Process for Remediation of Contaminated Soil" Presented at the 1996 *Midwest Thermodynamics and Statistical Mechanics Conference*, May 3-4, 1996, Madison, WI.
13. The Superfund Innovative Technology Evaluation Program: *Technology Profiles* 1992, 5th ed., EPA/540/R-92/077, pp. 260-261.
14. Ekhtera, M. R.; Mansoori, G.A.; Mensinger, M.C.; Rehmat, A. "Scale-Up of a Supercritical Fluid Extraction Process" Presented at the *American Institute of Chemical Engineers Annual Meeting*, November 10-15, 1996, Chicago, IL.
15. Test Methods For Evaluating Solid Waste, *U.S. EPA* 1986, SW-846, 3rd ed., vol. I-II.

Author Index

Affiliation Index

Forschungszentrum Karlsruhe GmbH, 270
Georgia Institute of Technology, 37
Industrial Research Limited, 76
Institute of Gas Technology, 280
Instituto Nacional de Engenharia e Tecnologia Industrial, 101
Instituto Superior Tecnico, 101
Kimberly-Clark Corporation, 57
Kumamoto University, 119
Massachusetts Institute of Technology, 242
Procter & Gamble Company, 68
Sandia National Laboratories, 255
Seoul National University, 110
Sogang University, 110

Technical University of Denmark, 90, 154
Texas A&M University, 208
U.S. Environmental Protection Agency, 280
U.S. Food and Drug Administration, 101
Universidad National del Sur—Consejo National de Investigaciónes Cientificas y Tecnicas, 51
University of Denmark, 90
University of Dortmund, 171
University of Illinois at Chicago, 280
University of Maine, 2
University of South Florida, 188
University of Tulsa, 232
Virginia Polytechnic Institute and State University, 134

Subject Index

Acetic acid group, role in solubilities of coumarin derivatives in supercritical carbon dioxide, 118
Adsorption, role in polycyclic aromatic hydrocarbon extraction from contaminated soils, 177–182, 184–185
Adsorption isotherms, estimation using supercritical fluid chromatography, 194–196
Adsorption kinetics, estimation using supercritical fluid chromatography, 197, 202
Adsorptive separations, frontal analysis supercritical chromatography, 223–224
Alcohols, role in polycyclic aromatic hydrocarbon extraction from contaminated soils, 171–186
Alkanes, role in polycyclic aromatic hydrocarbon extraction from contaminated soils, 171–186
Ammodytes sp., *See* Sand eel
Apparent diffusivity, definition, 31
Aqueous environmental media, supercritical extraction, 210–214

Bulk density, role in extraction of sage and coriander seed using near-critical carbon dioxide, 76–88

C205ω3, *See* EPA
C226ω3, *See* DHA
Cahn–Hilliard theory of spinodal decomposition, description, 31
Camphor, supercritical dioxide extraction, 225, 227–228f
Cape marigold, analysis of natural products, 154–169
Carbon dioxide
advantages as supercritical fluid, 171–172
applications, 37–38
critical conditions, 281
importance in geological processes, 3–4
near critical, *See* Near-critical carbon dioxide
role in polymer behavior, 15–20
See also Supercritical carbon dioxide
Carbon dioxide flow rate, role in extraction of sage and coriander seed using near-critical carbon dioxide, 76–88

Retention time, role in catalytic
supercritical water oxidation of
pyridine, 234–235, 236*f*
Rosemary, use of components, 101
Rosemary volatile compound extraction
using supercritical carbon dioxide
experimental apparatus, 102*f*, 103
experimental materials, 103
experimental procedure, 103–104
modeling, 104–105
particle size effect, 105, 106*f*, 107*t*
pressure effect, 105–106, 108–109
temperature effect, 105, 106*f*, 108–109

Sage, extraction using near-critical carbon
dioxide, 76–88
Sample injection, supercritical fluid
chromatography, 144*f*, 145, 147
Sample vacancy chromatography,
physicochemical property estimation, 203
Sand eel
analysis of natural products, 154–169
fish oil fatty acid ethyl ester
fractionation using supercritical
carbon dioxide, 90–99
Scattering intensity, definition, 31
Seeds, analysis using chromatographic
techniques, 154–169
Selective extractions, supercritical fluids,
224–227
Selectivity, supercritical fluids for
extraction, 71
Separation(s)
natural products, 68–74
use of supercritical fluids, 68
Separation selectivity, definition, 123
Silicon, fabrication problems, 257
Soil source, role in contaminated soil
remediation using supercritical fluid
extraction liquid-phase oxidation,
288–289*t*, 290
Solid environmental media, remediation
using supercritical fluids, 211, 215–219
Solid solubility estimation in supercritical
carbon dioxide
calculation of solid–supercritical fluid
equilibria, 39–41, 42*f*

Solid solubility estimation in supercritical
carbon dioxide—*Continued*
experimental procedure, 38–39, 42*f*
model development
coefficients with confidence limits, 47
correlation of solubility, 46*f*, 47–48
database of enhancement factors,
43, 44*t*
solvatochromic parameters for linear
solvation energy, 41, 43
Solid waste systems, efficient
management strategies, 57
Solubilities
coumarin derivatives in supercritical
carbon dioxide
acetic acid group effect, 118
disubstituted group effect, 114, 118
equilibrium condition effect, 114, 116*f*
experimental apparatus, 111, 113*f*
experimental description, 111
experimental materials, 111
experimental procedure, 111, 114
functional group effect, 114, 117*f*
position effect, 114, 117*f*
pressure effect, 114, 115*t*
previous studies, 111
structures of compounds, 111, 112*f*
temperature effect, 114, 115*t*
estimation using supercritical fluid
chromatography, 194
fish oil fatty acid ethyl ester
fractionation using supercritical
carbon dioxide, 93–94
role in polycyclic aromatic hydrocarbon
extraction from contaminated soils,
175, 177, 178*f*
See also Solid solubility estimation in
supercritical carbon dioxide
Solute solvatochromic parameters, solid
solubility estimation in supercritical
carbon dioxide, 37–48
Solvatochromic parameters for linear
solvation energy relationships
mathematical model, 41
solvent scales, 41, 43
Solvent composition, role in pressure-
induced phase separation, 18, 21–26

Highlights from ACS Books

T

1 Month